ATMOSPHERE IN SPACE CABINS
AND CLOSED ENVIRONMENTS

APPLETON·CENTURY·CROFTS   •   NEW YORK

DIVISION OF MEREDITH PUBLISHING COMPANY 1966

Atmosphere in Space Cabins and Closed Environments

edited by KARL KAMMERMEYER

State University of Iowa

ISBN-13: 978-1-4684-1374-8   e-ISBN-13: 978-1-4684-1372-4

DOI: 10.1007/978-1-4684-1372-4

F49460

CONTRIBUTORS

ARNOLDI, W. E., Hamilton Standard, Division of United Aircraft Corporation, Windsor Locks, Connecticut (76)

ASIMOV, ISAAC, Department of Biochemistry, Boston University School of Medicine, Boston, Massachusetts (249)

AUERBACH, ERIC E., Hamilton Standard, Division of United Aircraft Corporation, Windsor Locks, Connecticut (145)

CARL, W. R., Department of Chemistry, Purdue University, Lafayette, Indiana (223)

COTTON, J. E., Aero-Space Division, The Boeing Company, Seattle Washington (171)

FOSBERG, T. M., Aero-Space Division, The Boeing Company, Seattle, Washington (171)

FOSTER, JOHN F., Battelle Memorial Institute, Columbus, Ohio (104)

GRAF, G., Department of Chemistry, Youngstown University, Youngstown, Ohio (223)

HOAGLAND, R. E., Department of Chemistry, University of Cincinnati, Cincinnati, Ohio (223)

KAMMERMEYER, KARL, Department of Chemical Engineering, State University of Iowa, Iowa City, Iowa (1)

KUROWSKY, S. R., Department of Chemistry, Purdue University, Lafayette, Indiana (223)

MAJOR, COLEMAN J., Department of Chemical Engineering, The University of Akron, Akron, Ohio (120)

MILLER, R. L., Environmental Systems Branch, USAF School of Aero-Space Medicine, Brooks Air Force Base, Texas (186)

MONTEITH, L. E., Environmental Health Department, University of Washington, Seattle, Washington (171)

OLSON, R. L., Aero-Space Division, The Boeing Company, Seattle, Washington (171)

ROTH, EMANUEL M., The Lovelace Foundation for Medical Education and Research, Albuquerque, New Mexico (13)

RUSSELL, SID, Hamilton Standard, Division of United Aircraft Corporation, Windsor Locks, Connecticut (145)

SOLLAMI, BLASE J., Pioneer-Central Division, The Bendix Corporation, Davenport, Iowa (32)

TOCK, RICHARD W., Department of Chemical Engineering, State University of Iowa, Iowa City, Iowa (120)

WARD, C. H. Environmental Systems Branch, USAF School of Aero-Space Medicine, Brooks Air Force Base, Texas (186)

PREFACE

MANNED SPACE FLIGHT introduces into space travel parameters that are unique. Man can live without food for a reasonably long period; without water, the period becomes quite a bit shorter; but without air, the result—almost instantaneous—is death. This would make the atmosphere the most important consideration. In fact, however, man needs all three components: oxygen, water, and food; and if any one of them fails, he is doomed.

With our space efforts approaching trips of several weeks in length and certainly heading for month-long journeys, it is most appropriate to ask: Are we ready to provide an adequate atmospheric milieu for the astronauts?

The present volume represents the first integrated attempt to answer this question on a scientific level and on a broad basis of physical and mechanical, biological, biochemical and medical factors. The main features of this work were presented at a symposium of the Division of Industrial and Engineering Chemistry of the American Chemical Society, held in Atlantic City on September 13, 1965. The volume is an expanded and reorganized treatise based on, but not merely proceeding from, the symposium.

Obviously, medical aspects are of paramount importance. A down-to-earth appraisal of the status quo, presented by Dr. E. M. Roth, shows that the problem of 100% oxygen atmosphere still is beset with some uncertainty in the 200 to 500 mm. total cabin pressure range. Additionally, attention is called to inherent dangers of greatly increased flammability in such an atmosphere. Consequently, future developments in mixed atmospheres are likely to receive considerable attention.

A subject that concerns the general operation of atmosphere supply, as well as fuel supply for fuel cells, is that of gas storage. B J. Sollami presents an up-to-date analysis of supercritical and subcritical storage schemes. An actual design example concerned with a lunar storage requirement helps elucidate the interrelationship of the many interlocking factors.

That two papers deal with contaminant control is a sign that we are coming to grips with this very mundane, but extremely vital, problem. The essential features of the proposed solutions understandably overlap to a degree, as the parameters are set by the problem itself. Nevertheless, there is a significant difference: Auerbach and Russell attack the problem from the viewpoint of appropriate elimination and conversion of identifiable contaminants to easily removable components, whereas Cotton et al. use a strategic approach to try to control the materials going into a spacecraft so as to prevent the formation of contaminants. Obviously, a combination of these efforts should prove successful.

The management of carbon dioxide removal is treated in three contributions. Major and Tock suggest a mechanical separation process, a method not in a competitive position with presently existing adsorption type processes, that yet deserves attention for long-duration flights. The other two schemes are biological. G. Graf presents recent results on the use of an enzymatic CO_2 conversion system. This work is still largely in the experimental stage. Miller and Ward, who have carried out an exhaustive study of algae systems, present a monumental review and assessment of the suitability of such systems, concluding that algae systems, at their present stage of development, are not promising for a spacecraft proper. Such limitations, however, would not carry over to a potential use in large orbiting spacecrafts or to planetary colonies.

Oxygen recovery from carbon dioxide is a problem that will have to be solved if space flights of long duration are to become possible. John Foster covers the present situation in a review with a critical analysis of comparative merits of the technologically more advanced processes. He also introduces the possibility of a bioregenerative system based on a bacterial conversion scheme. Walter Arnoldi describes a system where a molten halide salt electrolysis cycle serves for the simultaneous removal of CO_2 from the atmosphere and conversion to free oxygen that can be returned to the cabin.

Finally, a fascinating, thought-provoking treatise by Isaac Asimov takes us forward in time to a point where all of these nagging problems have been solved, so that we can enjoy the ride.

Karl Kammermeyer

CONTENTS

SPACE TECHNOLOGY—
TODAY'S CHALLENGE TO SCIENCE

KARL KAMMERMEYER

Department of Chemical Engineering
State University of Iowa, Iowa City

We are living in the Space Age. Man's eternal desire to unravel the unknown provides the stimulus for the conquest of the universe. Early explorers of our planet have exposed themselves to untold hardships and dangers, but never before has such an "impersonally hostile" environment as that of space been challenged (Dreher, 1963).

Space travel is expensive. It will ultimately be more expensive than any program ever undertaken. The 1965 budget of the National Aeronautics and Space Administration exceeded 5 billion dollars and it will surely reach even larger proportions. Five billion dollars seems like a stupendous amount, and of course it is. However, if you consider that yearly expenditures in the United States for exploration of gas and oil sources approximate 4 billion dollars, and that 1964 foreign aid amounted to some 3.5 billion dollars, the NASA budget, by contrast, takes on more reasonable dimensions.

GENERAL CONSIDERATIONS RELATING TO SPACE FLIGHT

Space flight activities present many spectacular operations. Inevitably, the eye-catching and thrilling experience of a rocket launching centers attention on the showier aspects of such an undertaking. This very natural situation leads to a tendency to underestimate the importance of associated activities that may actually be the most critical ones in the ultimate success of *manned* space flight. The bioastronautical and life-support problems must be solved to assure the absolute safety and well-being of the astronaut (Konecci, 1963; Voris, 1963). This means that *a man must be provided with all essentials to sustain life in a completely synthetic, nonterrestrial environment*, or manned space flight for extended periods is out of the question.

There are numerous pertinent references (Eckman and Helvey,

1

1963; Grimwood, 1963; Inst. Aerospace Sci. NASA, 1962; Kammer-meyer, 1965; Koelle, 1963; Morris, 1965; Williams *et al.*, 1963), in particular, the extensive review in *Chemical and Engineering News* (1963). This account, which at times reminds us that science fiction has become fact, is a highly stimulating presentation.

For purposes of orientation, we should consider the present status of manned space flight. Summaries of United States and Soviet Union space flights as of December, 1965, are given in Tables 1 and 2.

TABLE 1

UNITED STATES MANNED SPACE FLIGHTS

	Orbits	Date	Astronaut	Time at 0 g (approx.)
MR-3[a]	SO	May, 1961	Alan B. Shepard, Jr.	5 min.
MR-4	SO	July, 1961	Virgil I. Grissom	5 min.
MA-6	3	Feb., 1962	John H. Glenn, Jr.	4.5 hr.
MA-7	3	May, 1962	M. Scott Carpenter	4.5 hr.
MA-8	6	Oct., 1962	Walter M. Schirra, Jr.	9.2 hr.
MA-9	22	May, 1963	Leroy Gordon Cooper, Jr.	34 hr.
GT-3	3	March, 1965	Virgil I. Grissom John W. Young	4.5 hr.
GT-4	62	June, 1965	James A. McDivitt ·Edward H. White II	97.5 hr.
GT-5	120	Aug., 1965	L. Gordon Cooper, Jr. Charles Conrad, Jr.	190.5 hr.
GT-6[b]	17	Dec., 1965	Walter M. Schirra, Jr. Thomas Stafford	25.5 hr.
GT-7[b]	220	Dec., 1965	Frank Borman James A. Lovell	330.5 hr.
GT-8[c]	7	March, 1966	Neil Armstrong David Scott	13 hr.

[a] Abbreviations: MR, Mercury Redstone; SO, Suborbital; MA, Mercury Atlas; GT, Gemini Titan.

[b] GT-6 and GT-7 were twin flights with rendezvous; closest approach about 1 ft.

[c] GT-8 performed successful docking maneuver with Agena.

EXTENT OF PROBLEMS

The multiplicity of problems generated by space flight are shown, in a condensed version, in Table 3.

The order in which the categories are listed is an indication of

TABLE 2

SOVIET UNION MANNED SPACE FLIGHTS

	Orbits	Date	Astronaut	Time approx. at 0 g
Vostok: I	1	April, 1961	Yuri A. Gagarin	1.5 hr.
II	17	Aug., 1961	Gherman S. Titov	25 hr.
III	64	Aug., 1962	Andrian G. Nikolayev	94.5 hr.
IV	48	Aug., 1962	Pavel R. Popovich	71 hr.
V	82	June, 1963	Valery F. Bykovsky	119 hr.
VI	49	June, 1963	Valentina Tereshkova	71 hr.
Voskhod: I	16	Oct., 1964	Vladimir Komarov Konstantine P. Feoktistov Dr. Boris B. Yegorov	24 hr.
II	18	March, 1965	Pavel I. Belyayev Alexei Leonov	25.5 hr.

All were orbital flights; no suborbital flights have been announced.
Vostok III and IV were twin flights; closest approach, 3.5 miles.
Vostok V and VI were twin flights; closest approach, 3 miles.

the approximate magnitude of the funds that are being expended for the respective activities.

To cover any one of these subjects with a reasonable degree of completeness would require an encyclopedia; available documentation is simply overwhelming.

In addition to the great many current references, a growing number of books are concerned with the problems of man in space (Bourne, 1963; Brown, 1963; Gerathewohl, 1963; Schaefer, 1964; Wunder, 1965, 1966). While the references quoted here are quite recent, the rate at which publications appear makes it impossible to have a paper go into print without being already somewhat obsolete in regard to current events.

TABLE 3

SPACE FLIGHT ACTIVITY CATEGORIES

1. Propulsion systems and fuels for launching, flight control, and reentry
2. Material science problems created by new technological demands
3. Power supply systems within the spaceship
4. Environmental requirements for manned space flight
5. Impact on education

ENVIRONMENTAL AND LIFE-SUPPORT REQUIREMENTS

Dr. Hubertus Strughold of the United States Air Force Aerospace Medical Center (Brooks Air Force Base, Texas) has been concerned for many years with all aspects of man in space. I quote his summing up of environmental conditions in space in order to emphasize the multitude, severity, and complexity of the problems created by man in space (Gill and Gerathewohl, 1964; Strughold, 1961; USAF School of Aerospace Med., 1963-1964). Thus, space means: "... no oxygen for respiration, no physiologically effective barometric pressure; no air available for pressurization of a cabin; occurrence of high energy particles of solar and cosmic origin in their original, primary form; the full range of the solar electromagnetic spectrum from 6 Ångström units to radio waves of more than 10 meters; no indirect sunlight, but intensive direct sunlight; heat transfer only by radiation; no sound propagation, no aerodynamic navigation and no air resistance" (Strughold, 1961). The medical problems posed by this situation are aptly treated in a publication by NASA's Dr. F. B. Voris, who asks, "Can man survive in space?" (Voris, 1963), and by Dr. S. P. Vinograd in a summary of medical experiments of the Gemini program (Vinograd, 1964).

Evidently, intelligent participation of the scientist in the man-problems of space flight will require branching out into multi-disciplinary sciences. Physiological, biomedical, and biochemical knowledge will have to be acquired (Bourne, 1963; Brown, 1963; Konecci, 1963; Siegel *et al.,* 1961, 1962, 1963; USAF School of Aerospace Med., 1963-1964) to evaluate all possible solutions, including the attractive possibilities of a biological system. The immediate problems, however, are so pressing that a biological system of life support remain a matter for future development. Current problems must be solved by mechanical-chemical means, because these are the only ones sufficiently far advanced to be useful.

The situation is one wherein a living organism, an independently functioning entity that developed on earth and is entirely dependent in all of its physiological functions upon its natural environment, is removed from all supporting surroundings and placed in a capsule with the stark nothingness of space around him. He is deprived of his natural milieu, and we must make up for this. It is as simple as that.

Basic requirements

The basic necessities that must be provided to keep man alive in any surroundings are: *Oxygen, water, and food* (Keating and Rounding, 1961). As long as there is life, the body's functions continue unceasingly and we must supply means to maintain breathing,

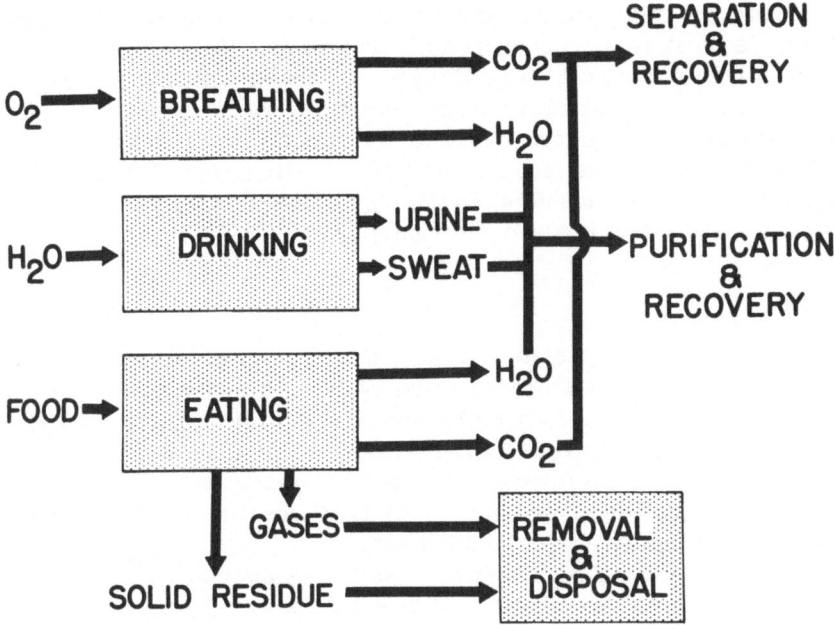

Figure 1. Diagram of life-support problems.

drinking, eating, and waste elimination. Consequently, oxygen, water, and food are basic and critical factors. Obviously, if any one of them, or more than one, could be recycled, the critical factors would be reduced accordingly.

Figure 1 points up the essential features of the necessary life-support system. On this basis we can consider the requirements in the following categories: (1) maintenance of breathable atmosphere; (2) water cycle; (3) food cycle. With all these objectives, attention should be called to some very important aspects, listed in Table 4.

Every process must be developed so that it will work under conditions of weightlessness; weight considerations and power requirements are extremely important; and reliability demands will have a decisive influence on final selection.

TABLE 4

PROCESS CRITERIA

1. Balance of power requirements against weight
2. Suitability of equipment for the confined facilities in a spacecraft
3. Reliability and attainable degree of automation
4. Operability under conditions of weightlessness
5. Quality of product

How does the situation appear in regard to possible *recycling*? Water is most easily recovered and recycled and this problem has been solved in several ways. Oxygen, separable as carbon dioxide, presents considerable difficulties. The recovery of oxygen from carbon dioxide may, in fact, become the pivotal problem upon which will hinge the ultimate success of long-duration manned flight. Recycling of food by means other than biological conversion is as yet beyond the capabilities of our technology.

Food

For a long time to come, all food will have to be carried, and the unique conditions of storage, preservation, and of ingestion under weightlessness require careful planning (Nanz, 1964). A rather comprehensive study was reported by Miller and Halpert (1962) on a food refrigeration system for spacecrafts. The *logistic analysis* indicated that for a three-man mission of 30 days' duration, a refrigerated volume of 16 cu. ft. would be adequate.

Of course, food consumption results in *human waste production* and thus creates the problem of waste *disposal*. For space flights within the foreseeable future, the waste material, that is, fecal matter, will have to be stored. Such storage must be either above 120°C. or below 0°C. Miller's report covers consideration of the storage of fecal material in the freezer compartment of the refrigerator. As food is consumed, more than enough space becomes available to store the waste material in properly packaged condition.

Our knowledge of human waste, its composition, and how to handle it other than in a biological (that is, natural) system, is woefully inadequate. A study on fecal composition has been started by Goldblith and his co-workers (Goldblith and Wick, 1961). The information is not only very valuable, but it also serves the useful purpose of stressing the magnitude of the problems to be solved, especially for long-duration flights.

The possibility of utilizing the waste products in a *bioregenerative subsystem* would become of definite interest when power and space are less critical than at present (Casey and Lubitz, 1963; Lachance and Vanderveen, 1963). Thus, an orbiting space station and certainly colonies on the moon or on the planets are the most likely places where biological systems will be established.

Cabin atmosphere

The essential requirements in cabin atmosphere maintenance are adequate oxygen concentration, carbon dioxide removal, and elimination of toxic and odor-contributing substances (Scano, 1962).

Oxygen

The situation concerning desirable oxygen concentration is still unresolved (Eckman and Helvey, 1963; Strumza, 1962). The decision by NASA, now in effect, to use a 100% oxygen atmosphere for the moon flight has met with considerable opposition from some medical and physiological workers. The consensus of the opposition is that the testing program has been inadequate, and that the problem of *oxygen toxicity* is a serious one.

Oxygen supply

The relatively short-time orbital flights that have been made so far did not pose any difficult problems. In the United States Mercury flights, oxygen was carried *at cabin temperature* as a compressed gas at 7,500 p.s.i. (510 atm.) and the oxygen concentration was replenished by sensing-instrument control (Inst. Aerospace Sci. NASA, 1962; Williams *et al.*, 1963). Apparently, the oxygen in the Soviet flights was largely obtained from superoxides (Volynkin *et al.*, 1962) through the reaction with the CO_2 in the cabin air. In the Gemini flights, supercritical storage of oxygen has been used and is contemplated for future excursions. The Apollo program also calls for supercritical storage of oxygen for both life support and fuel cell operation. Flights of longer duration will most likely depend on the development of a recycling process that requires recovery of oxygen from CO_2 .

Contaminants

Toxic gases and *odor-substances* are normally present only in small amounts and represent both a nuisance and a potential hazard in long-time exposure. Their removal is accomplished effectively by adsorbents, such as activated carbon; some compounds will require chemical absorbents. Of course, the long-duration flights will require some provision for regeneration of the adsorbent, or the development of after-burners with subsequent adsorption and/or absorption of reaction products. A concise summary of chemical problems, including the matter of trace contaminants was presented by Thompson (1965).

Carbon dioxide

From the standpoint of relative importance, the removal of *carbon dioxide* ranks first. The concentration of CO_2 in normal air is usually taken as 0.03% by volume. The air exhaled by the human body contains approximately 4.5% of CO_2 by volume, and there is an appreciable build-up of CO_2 in any enclosed space. Past experience indicates that the carbon dioxide concentration should not be permitted to exceed 0.5% by volume, based on a total pressure of 760 mm. Thus the partial pressure of CO_2 should remain below 3.8 mm. mercury. While the initial Mercury flights used a mixed-gas atmosphere (that is, nitrogen and oxygen), the later Mercury flights used pure oxygen. The Gemini and Apollo flights are scheduled to use a 100% oxygen atmosphere at about 5 p.s.i. absolute pressure (1/3 atm.) with the maximum partial pressure of CO_2 at 2 mm. (USAF School of Aerospace Med., 1963-1964).

Methods for CO_2 removal

The most promising methods for removing the CO_2 generated in the cabin of a spaceship are listed in Table 5.

Algae systems are certainly in the future but will require a great deal of developmental work. Gaseous diffusion systems are receiving active attention. The inherent power requirements make such a system unattractive at present, but in combination with other

TABLE 5

POSSIBILITIES FOR CO_2 REMOVAL

Method	Comment
Algae systems	Long-range development
Gaseous diffusion (use of selective permeable membranes)	Power and area requirements still appear excessive
Absorption (chemical absorbents, Li_2O or LiOH, etc.)	For short flights, the best available method
Regenerable adsorbents	Suitable for medium-duration flights with space vacuum for regeneration of adsorbent
Regenerable absorption systems	Long-range development

Adsorbents that can be regenerated under *space cabin* conditions and ultimately *without* use of space vacuum will become of extreme importance.

methods this "single-phase operation" looks promising. The absorption systems also would include the use of superoxides, such as potassium superoxide (KO_2) and enzyme suspensions. Allen (1964) evaluated absorption methods in a comprehensive report.

A great deal of work has already been done with *regenerable adsorbents* (DelDuca *et al.*, 1962; Major *et al.*, 1965). Extensive cycling tests with alternate adsorption and desorption are an essential aspect of testing because the effects of potential contaminants in the atmosphere may not become evident in short-time evaluation.

Regeneration of CO_2

The recovery of oxygen in adequate purity from CO_2 has so far defied completely successful solution, although a number of systems look promising.

OBJECTIVE OF SYMPOSIUM

The symposium presentations are intended to answer the question: "Where do we stand today in regard to space cabin atmosphere?" Thus, the over-all view and the medical aspects delineate the extent of the problems. The matter of gas storage is also of general importance. It is certainly significant that two of the discussions deal with the subject of contaminant control. Likewise, two contributions treat the serious and most important area of oxygen recovery from carbon dioxide. Also, the matter of biological atmosphere regeneration is presented by two different approaches. The removal of CO_2 by present conventional methods of molecular sieve adsorption is not discussed, as there is adequate information available from a number of sources. However, the potential use of gaseous diffusion cells deserves some attention, even though such cells are not of immediate importance because of high energy requirements.

Finally, even as we acknowledge the magnitude and complexity of the problems involved in launching a spacecraft, placing it in orbit, or sending it on a space mission and maneuvering it to successful reentry, we must, without detracting from the importance of the operations just mentioned, continue to recognize that the astronaut is still the most important system component in manned space flight.

REFERENCES

Allen, J. P. (1964). Carbon dioxide management. I. Technique for carbon dioxide absorber evaluation. W-PAFB FDL TDR 64-67, P. I.

Bourne, G. H., ed. (1963). Medical and biological problems of space flight. Academic Press, New York.

Brown, J. H. U. (1963). Physiology of man in space. Academic Press, New York.

Casey, R. P., and J. A. Lubitz. (1963). Algae as food for space travel, a review. Food Technol., 17 (11): 48-56.

Chem. and Eng. News (1963). The U.S. effort in space. Special Rept., Chem. and Eng. News, 41 (38): 98-128 (Sept. 23); (39): 70-100 (Sept. 30).

DelDuca, M. G., R. C. Huebacher, and A. E. Robertson (1962). Regenerative environmental control systems for manned earth-lunar spacecraft. TAPCO publication (Division of Thompson Ramo Wooldridge, Inc.).

Dreher, C. (1963). Martyrs on the moon. Harper's Mag., 226 (1354): 33-38 (March).

Eckman, P. K., and W. Helvey (1963). Spacecraft and life system. Astronaut. Aerospace Eng., 1 (10): 20-28.

Gerathewohl, S. J. (1963). Principles of bioastronautics. Prentice-Hall, Englewood Cliffs, N. J. Also (1965): Man's role in space. Intern. Sci. Tech., 45: 64-74.

Gill, Jocelyn R., and S. J. Gerathewohl (1964). The Gemini science program. Astronaut. Aeronaut., 3 (11): 58-65.

Goldblith, A., and Emily L. Wick (1961). Analysis of human fecal components and study of methods for their recovery in space systems. ASD Tech. Rept. 61-419 (August), W-PAFB.

Grimwood, J. M. (1963). Project Mercury, a chronology. NASA SP-4001; MSC Publ. HR-1.

Inst. Aerospace Sci. NASA (1962). Proc. Natl. Meeting on Manned Space Flight, St. Louis (April 30-May 2).

Kammermeyer, K. (1965). Regenerative methods in closed biological systems. Raumfahrtforschung, 9(2): 91-95 (April-June). In German).

Keating, D. A., and R. W. Roundy (1961). Closed ecology. WADD TR 61-129.

Koelle, H. H. (1963). Manned planetary flight—where are we today? Astronaut. Aerospace Eng., 1 (10): 131-133.

Konecci, E. B. (1963). Bioastronautics review—1963. Presented at XIV Intern. Astronaut. Federation Meeting, Paris (Sept. 26-Oct. 1). NASA Headquarters, Washington, D. C.

Lachance, P. A., and J. E. Vanderveen (1963). Problems in space foods and nutrition: foods for extended space travel and habitation. Food Technol., 17 (5): 59-64.

Major, C. J., B. J. Sollami, and K. Kammermeyer (1965). Carbon dioxide removal from air by adsorbents. Ind. Eng. Chem. Process Design Develop., 4: 327.

Miller, R. A., and S. Halpert (1962). A food refrigeration and habitable atmosphere control system for space vehicles, design, fabrication, and test phases. AMRL-TDR-62-149 (Dec.); General Electric Co., under Contract No. AF 33 (616)-6902.

Morris, D. N. (1965). Third manned space flight meeting. Astronaut. Aeronaut., 3 (4): 76-80.

Nanz, R. A. (1964). Food in flight. Space World A-3 (Jan.): 12-14.

Scano, A. (1962). Major biological problems related to environmental situations of space flight, with particular reference to air regeneration. Proc. Intern. Congr. Man and Technol. in the Nuclear and Space Age, Milan (Apr. 18-21), pp. 519-533.

Schaefer, K. E., ed. (1964). Bioastronautics. Macmillan, New York.

Siegel, S. M., et al. (1961). Effects of reduced oxygen tension on vascular plants. Physiol. Plantarum, 14: 554-557; 15: 304-314; 437-444 (1962); 16: 549-555 (1963).

_____, L. A. Halpern, C. Giumarro, G. Renwick, and G. Davis (1963). Martian biology: the experimentalist's approach. Nature, 197: 329-331.

_____, G. Renwick, and L. A. Rosen (1962). Formation of carbon monoxide during seed germination and seedling growth. Science, 137: 683.

Strughold, H. (1961). Extraterrestrial environments—bioastronautical aspect. Lectures, Inst. Technol., USAF Air University, W-PAFB (March).

Strumza, M. V. (1962). The usefulness of an inert gas in an artificial environment. Intern. Congr. Man and Technol. in the Nuclear and Space Age, Milan (Apr. 18-21), pp. 487-489.

Thompson, F. W., Jr. (1965). Chemical problems in atmosphere control. ASME 65-AV-47 (March).

USAF School of Aerospace Med., (1963-1964). Aerospace Med. Div. Lectures in aerospace medicine. Brooks Air Force Base, Texas, Feb. 4-8, 1963; Feb. 1964.

Vinograd, S. P. (1964). Medical experiments in Gemini. Astronaut. Aeronaut., 2 (11): 70-73.

Volynkin, Y. M., V. I. Yazdoviskiy, et al. (1962). The first manned space flights. In English, Translation FTD-TT-62-1619 (Air Force Systems Command, W-PAFB, Ohio).

Voris, F. B. (1963). Medical aspects of space flight. Space World, A-2 (Nov.-Dec.): 5-9.

Williams, W. C., *et al.* (1963). Mercury project summary, including results of the fourth manned orbital flight, May 15 and 16, 1963. NASA SP-45.

Wunder, C. C. (1965). In Hypodynamics and hypographics; the physiology of inactivity and weightlessness (M. McCally, ed.). Academic Press, New York.

_____(1966). Life into space. F. A. Davis CO., Philadelphia (in press).

2

MEDICAL CONSIDERATIONS IN THE SELECTION OF SPACE CABIN ATMOSPHERES

EMANUEL M. ROTH

The Lovelace Foundation
for Medical Education and Research, Albuquerque

In the past, the pressure of engineering commitments involved in the development of spacecraft required that decisions on space cabin atmospheres be made early, often before the physiological tolerance to unnatural gaseous environments could be determined. Selection has been based primarily on engineering grounds with the burden of proof on the physiologist that the proposed environments could not be tolerated. While this approach has been adequate in the past, it will have to be modified for the longer and more hazardous missions projected for the future.

The physical variables of the space cabin environment that must be considered are shown in Table 1. There are also many physiological and pathological considerations with which these environmental variables may interact (Table 2). The interactions indicated are complex, and many are beyond the scope of this symposium. This discussion, therefore, is limited to interactions that are of particular interest to the chemical engineer.

To satisfy the cabin designer, one should keep the total pressure in the cabin as low as possible. Analysis of the weight penalty for different total pressures in the cabin suggests that below 5 to

TABLE 1

CABIN ENVIRONMENT VARIABLES

1. Total pressure	7. Thermal properties of gas
2. Oxygen pressure	8. Circulation of gas
3. Carbon dioxide pressure	9. Temperature of gas
4. Inert gas pressure	10. Leakage rate of gas
5. Gaseous trace contaminants	11. Duration of exposure
6. Water vapor pressure	12. Gravitation level

TABLE 2

PHYSIOLOGICAL AND PATHOLOGICAL CONSIDERATIONS

1. Alertness and performance
2. Communication
3. Time of useful function
4. Decompression syndromes
   (a) Aeroembolism and bends
   (b) Barootitis and barosinusitis
   (c) Cardiovascular collapse
5. Respiratory physiology
   (a) Atelectasis
   (b) Hypoxia
   (c) Hypo- and hypercapnia
   (d) Hemoglobin control
6. Oxygen toxicity syndrome
7. Radiation sensitivity
8. Fire and blast hazards
   (a) Meteoroid penetration effects
   (b) Cabin fire control
9. Bacterial flora changes and infections
10. Water physiology
11. Thermal control problems

7 p.s.i., the weight of a cabin wall is independent of the pressure. Factors other than hoop tension determine the weight. Above 7 p.s.i., the weight penalty rises steeply with increasing pressure.

The selection of oxygen percentage is strongly determined by the total pressure. Figure 1 indicates the constraints of the pressure-oxygen system.

The line marked "sea level equivalent" represents those pressures and oxygen percentages equivalent to the partial pressure of 20% O_2 at sea level. Pressure must be reduced to 33,000 feet or to less than 4 p.s.i. in order to maintain the same partial pressure with 100% O_2. The heavy lines enclose the zone of unimpaired function. Hypoxia is limiting on the lower border and oxygen toxicity on the upper border. Acclimatization to hypoxia allows survival at the lower and left borders of the curve. There is no evidence at present for the development of tolerance to high pressures of oxygen. There is great individual variability to oxygen toxicity. The highest oxygen tolerable in all subjects for long periods of time seems to follow the upper fine line of Figure 1. However, in recent years evidence has accumulated that in the zone of 80 to 100% O_2 (hatched area) there may well be more severe time limits than previously expected (Roth, 1964b).

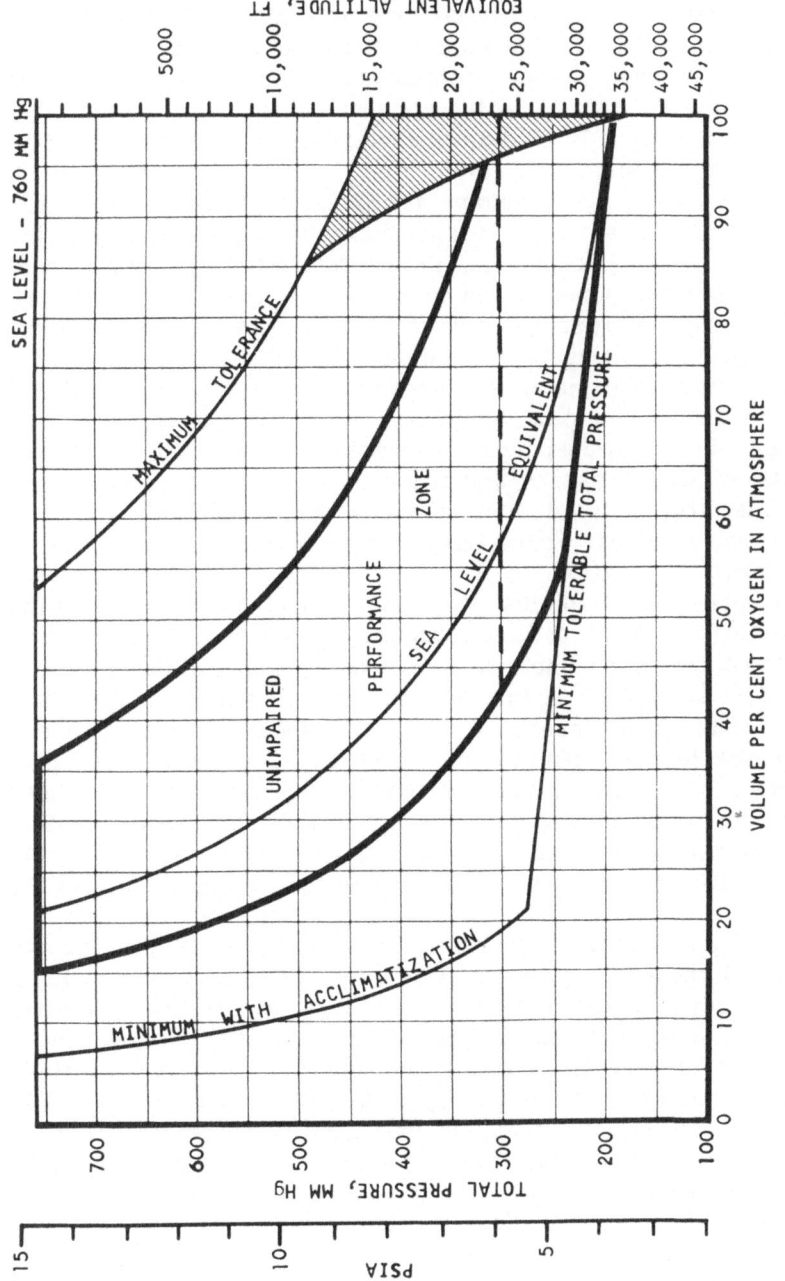

Figure 1. Oxygen-pressure effects; adapted from Luft (1962).

Figure 2 indicates the time and pressure dependence of oxygen toxicity. The ordinate is partial pressure of O_2. The abscissa is time to first symptoms. At high pressures of several millimeters Hg, symptoms occur within 10 minutes and these are mostly of the central nervous system. In the zone of 1 atm. to $\frac{1}{2}$ atm., the first symptoms are usually of the respiratory tract, such as bronchitis and pulmonary edema.

Below $\frac{1}{2}$ atm., the symptoms are variable. Atelectasis or alveolar collapse is seen in susceptible subjects, especially in the presence of high g loads. This is manifested by pains in the chest exaggerated by deep inspiration. In one 14-day experiment at 5 p.s.i., 100% O_2, subjects experienced an oxidative hemolytic anemia and excessive protein and casts in the urine suggestive of renal damage. Under similar circumstances, other subjects have gone as long as 30 days without symptoms. Symptoms may well be due to trace contaminants in the atmosphere. At present, the interaction between 100% O_2 at low pressures and trace contaminants of known toxic materials are being studied at the Aeromedical Laboratory at Wright-Patterson Air Force Base with technical aid from the Aerojet General Corporation.

In operational vehicles, we have reached the 100 hours of the White-McDivitt flight, $\boxtimes$, and 200 hours of the Cooper-Conrad flight, $\odot$, with no apparent symptoms but possible changes in the red blood cells. The asymptote at the abscissa of Figure 2 is therefore uncertain. The oxygen pressure of 5 p.s.i. (about 100 mm. Hg above normal sea-level oxygen pressure) used in Mercury and Gemini may represent borderline O_2 toxicity. Antioxidant defenses have been selected in biological systems (through evolution) for less than sea-level equivalents of oxygen pressure. Over long periods of time, the 100 mm. Hg excess pressure may lead to oxidation injury of the body.

The 5 p.s.i. pressure was used in Mercury and Gemini for two reasons. It reduced the danger of decompression sickness on going from sea level to cabin pressure, and it decreased the atelectatic tendency at lower pressures. Table 3 demonstrates the differences in collapse tendency in a lung filled with air, a lung filled with O_2 at 1 atm., and a lung filled with O_2 at 0.26 atm. There is a 370-fold difference between the first and the third. Nitrogen acts to brake the collapse, as would any of the inert gases. The lower the water or blood solubility of a gas, the less the collapse tendency.

It can be seen in Table 4 that helium and neon would be even more effective than nitrogen in preventing atelectasis. Fortunately, forced deep breathing, especially during takeoff and reentry and during other high g maneuvers, is also effective in reducing the tendency toward collapse of the alveoli.

Oxygen toxicity is of great concern to the radiobiologist. It

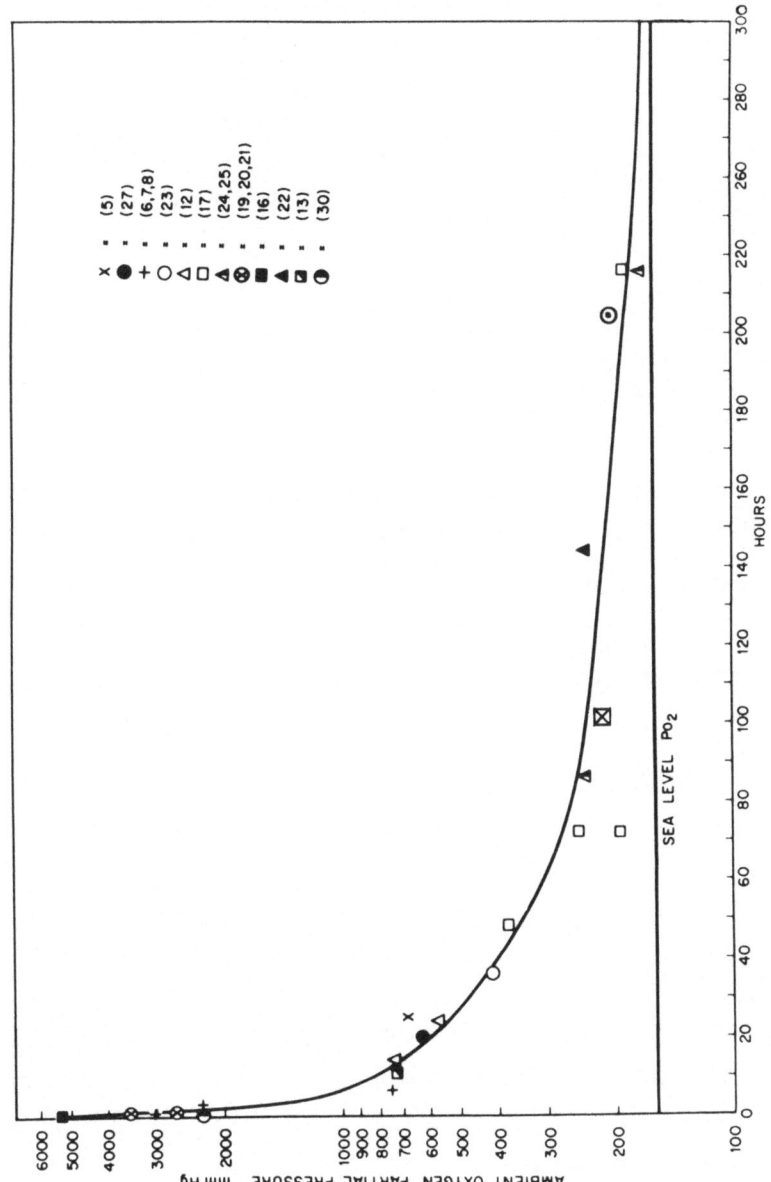

Figure 2. Time and pressure dependence of oxygen toxicity; adapted from Welch (1963).

TABLE 3

RATES OF COLLAPSE OF ONE LUNG [a]

	Rate of collapse		Time to collapse lung (min.)
	Observed cc. (min.)(kg.)	Relative	
Breathing:			
Air at 1 atm.	0.046	1	370 [b]
O_2 at 1 atm.	2.87	62	6
O_2 at 0.26 atm.	17.20 [b]	370	1 [b]

[a] With airway obstructed, when N_2 is present (breathing air) and when N_2 is absent (breathing O_2); after Dale and Rahn (1952).

[b] Calculated.

raises the question of whether or not the excess of about 100 mm. Hg above normal pO_2 can increase the radiation sensitivity of astronauts. Hyperbaric oxygen has been used in radiotherapy to increase the sensitivity of tumors to radiation. However, several atmospheres of oxygen pressure are required to produce a significant effect (Roth, 1964b).

Several years ago, studies at the Republic Aviation Corporation

TABLE 4

BIOCHEMICAL PROPERTIES OF INERT GASES [a]

	Gas					
	He	Ne	Ar	Kr	Xe	N_2
Bunsen solubility coefficient in:						
Water (38°C.)	0.0086	0.0097	0.026	0.045	0.085	0.013
Olive oil (38°C.)	0.015	0.019	0.14	0.43	1.7	0.061
Human fat (37°C.)	—	0.020	—	0.41	1.6	0.062
Oil: water solubility ratio	1.7	2.1	5.3	9.6	20.0	5.1
Relative diffusion through gelatin (23°C.)	1.0	(0.42)	0.30	0.21	0.13	0.36

[a] After Roth (1965).

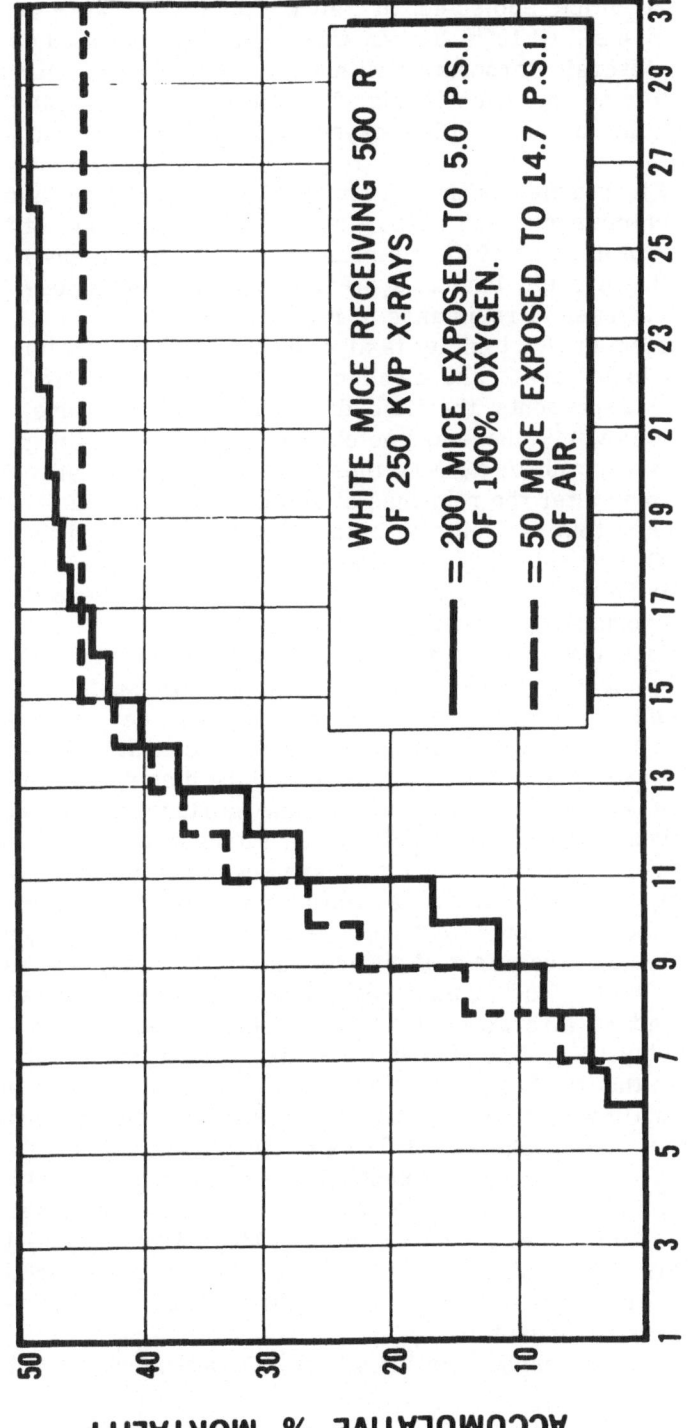

Figure 3. Effect of 5 p.s.i. 100% O_2 on radiation sensitivity of white mice receiving 500 r. of 250 k.v.p. X-rays. After Kelton and Kirby (1964).

suggested that mice subjected to 250 k.v.p. X-rays were more sensitive at 5 p.s.i. of 100% oxygen than in air. Recent studies at the Douglas Aircraft Corporation with a larger number of animals suggest there is no operationally significant difference in sensitivity produced by the 5 p.s.i., 100% oxygen environment (Kelton and Kirby, 1964).

Figure 3 is a review of these results. The slightly greater resistance of animals at 5 p.s.i. 100% oxygen may be a result of the atelectasis seen in some of the animals in the pure oxygen environment. Atelectasis could lead to systematic hypoxia and a possible paradoxical increase in resistance to radiation.

The problem of CO_2 toxicity is also present in a space vehicle. The data on toxic levels are much more clear than with oxygen.

Figure 4 represents the time dependence of toxic levels of CO_2 in man. Below $1\frac{1}{2}\%$ there are no changes in gross physiological function. However, above $\frac{1}{2}\%$ CO_2 there are biochemical adaptive changes that may alter the response of animals to radiation. There are no studies of this relationship in the literature. Current design limits for CO_2 are set at $\frac{1}{2}\%$. Emergency modes may take the CO_2 higher. It may be well to get data on changes in radiation sensitivity that may be produced by higher pressures of CO_2.

Another question of interest is the fire and blast hazard of pure oxygen environments. In a recent publication, the physical chemistry of fire and blast in numerous gaseous environments was reviewed. There are many variables involved. Let us look at just a few.

Figure 5 represents the minimum quenching distance and minimum ignition energy for different propane mixtures at different total pressures and with different oxygen-nitrogen compositions. Effects of several orders of magnitude are seen on going from 20% O_2 to 100% O_2. The data would be similar for hydrocarbon sprays and mists from disrupted hydraulic lines and coolant systems.

Figure 6 shows the effect of different gas mixtures on the rate of burning of cloth. The ordinate compares the burning rate with that of a standard aircraft cabin at 8,000 feet. The solid lines are lines of constant pO_2. The dotted lines are lines of constant $O_2\%$. It can be seen that the 5 p.s.i. 100% O_2 environment of Mercury and Gemini would allow cloth to burn 3.5 times faster than in a standard aircraft cabin. Atmospheres of 7 p.s.i. 50% O_2-50% N_2 recently suggested for Apollo and subsequent missions would allow cloth to burn 2 times as fast as in jet cabins. Fireproofing of material is also much less effective in 100% O_2 (Roth, 1964c). For present-day missions this fire hazard is acceptable. For manned orbital laboratories or other missions of high fire risk, one must use 100% O_2 with extreme caution.

An alternate atmosphere is 7 p.s.i. 50% O_2-50% inert gas. This

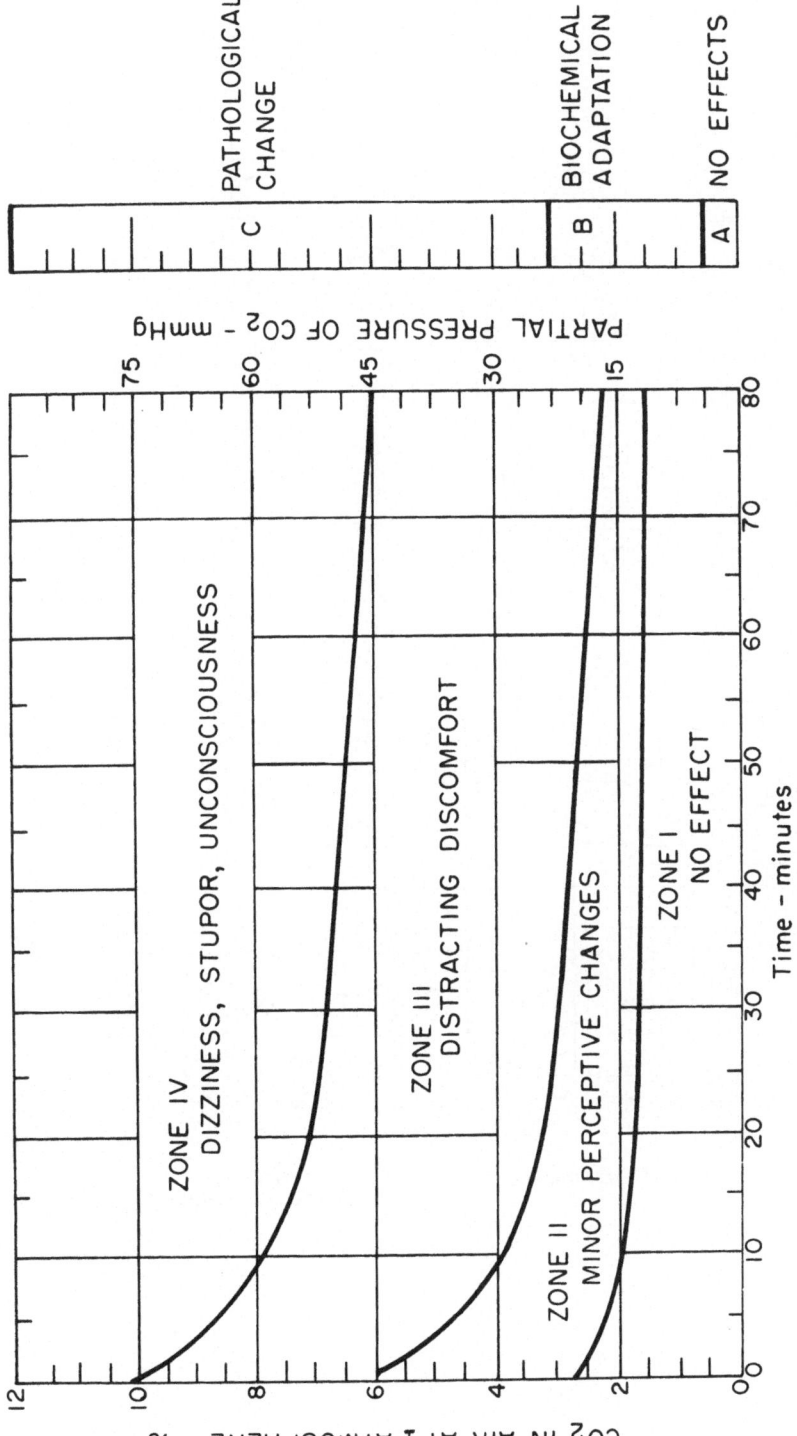

Figure 4. Time dependence of carbon dioxide toxocity; after Roth (1964a).

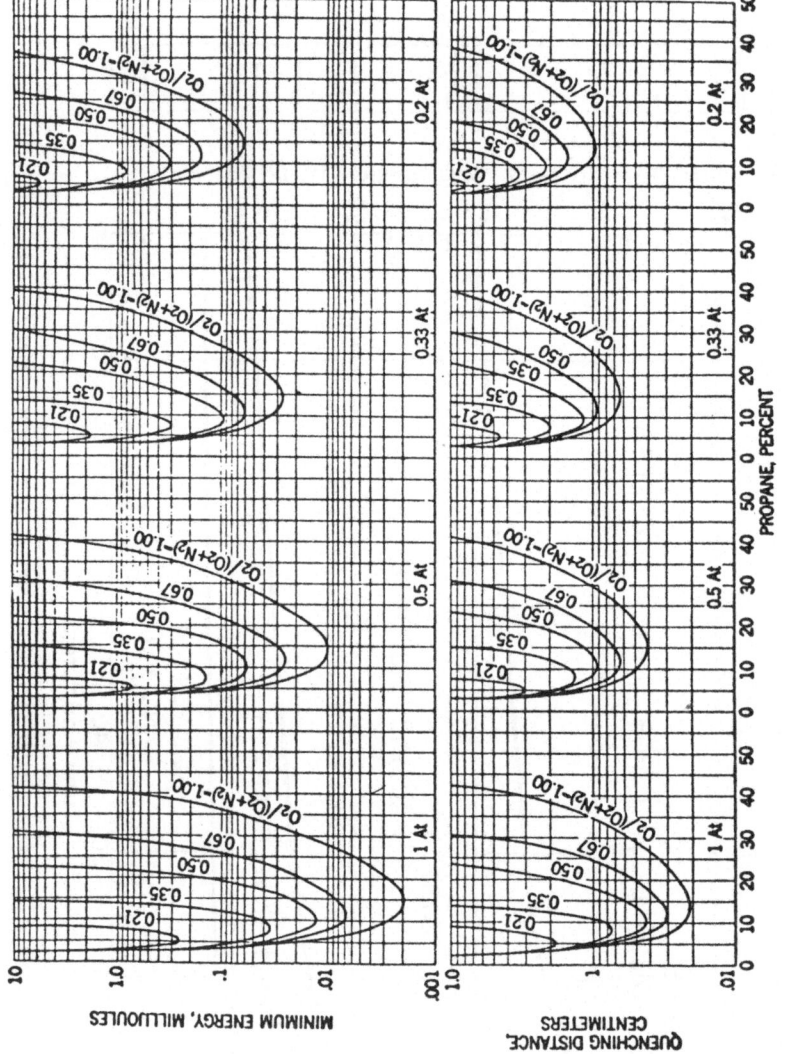

Figure 5. Ignition of propane in varied oxygen environments; after Lewis and von Elbe (1951). Abbreviation: At = atmosphere

TOTAL PRESSURE - MM. HG

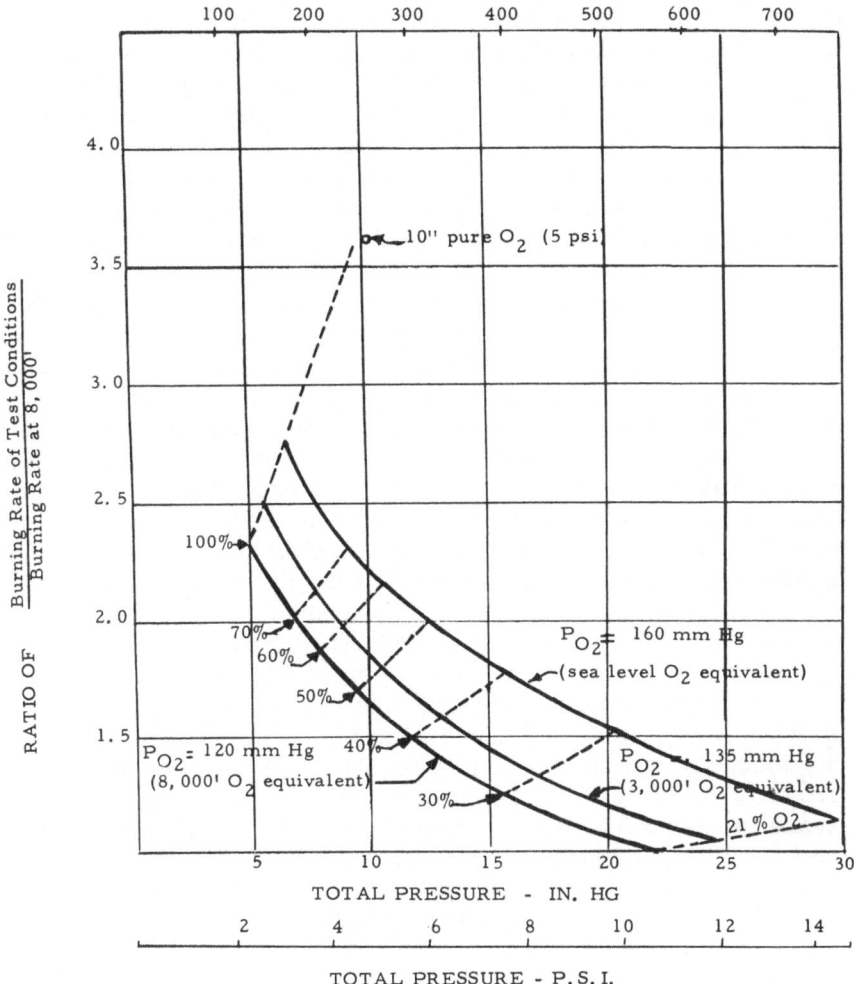

Figure 6. Burning rate of cloth under varied atmospheric conditions; adapted from Klein (1960).

approach decreases the fire hazard but brings into play two other factors: the difficulty of controlling a two-gas system, and the increased hazard of decompression sickness. Control of two-gas systems is much less reliable than the simple anaeroid control of 100% O$_2$ atmospheres. Only recently has there become available reliable, flyable O$_2$ sensors.

Decompression sickness arises from the release of inert gas

bubbles in the joint tissues and bloodstream produced by the decrease in ambient pressure. In decompression to altitude, the pressure must be decreased by a factor of more than 2 before symptoms arise. If a cabin at 7 p.s.i. of 50% O_2-50% N_2 were to decompress and the crew were to inflate their suits to the standard 3.5 p.s.i., there would be a small but definite chance of getting into trouble. In a recent paper it was calculated that less than 7% of a physically conditioned astronaut population should have mild symptoms of bends after severe exercise following such a decompression. At rest, less than 1% should get into trouble from bends (Roth, 1965).

How effective are the other inert gases in decreasing the symptom frequency or severity? This is a most complex problem. Unfortunately, there are not enough theoretical or empirical data available for a definitive analysis of this question. Bubble theory can give a degree of hazard for any given gas supersaturation ratio at a specific body site. The supersaturation ratios for the critical sites in the body are constantly changing as the inert gas is exchanged with the ambient environment in the lungs. The time constants of inert gas decay in the different body tissues are not known with enough accuracy to calculate the prevailing supersaturation ratios at any period after decompression.

By assuming a constant supersaturation ratio one can get only a first approximation of the relative hazard of each gas. The specific inert gas factors controlling the rate of growth of a bubble after decompression can be determined from the equation of state. The following equations (from Roth, 1965) indicate that the rate of growth of bubble at constant pressure or the rate of pressure increase at constant volume are determined only by the permeation coefficient $(\alpha_g D_g)$ of the gas in question when: A is the surface

$$\frac{dV_g}{dt} = \frac{A R T \alpha_g D_g \Delta P_g}{P \Delta x}$$

$$\frac{dP_g}{dt} = \frac{A R T \alpha_g D_g (\Delta P_g)}{V \Delta x}$$

area of a bubble, R the gas constant, T the absolute temperature, α_g the solubility coefficient of the gas, D_g the Fick diffusion coefficient of the gas, ΔP the pressure gradient between tissue and bubble, Δx the diffusion path length, P the pressure, and V the volume.

Using this relationship and the fact that the removal of gas from any tissue by the blood is limited not by diffusion, but by the perfusion of the tissue with blood, one can calculate a factor repre-

TABLE 5

INERT GAS BUBBLE FACTORS IN DECOMPRESSION SICKNESS[a]

Bubble site and factors	Gas					
	He	Ne	A	Kr	Xe	N_2
CASE 1 Adipose tissue (autochthonous)						
Bubble factor $\left(\dfrac{\alpha^2_{fat} D_{blood}}{\alpha_{blood}}\right)$	0.24	0.17	2.2	7.8	49.0	1.0
Relative bubble factor ($N_2 = 1$)	0.24	0.17	2.2	7.8	49.0	1.0
CASE 2 Adipose tissue (intravascular *in situ*) Phase 1. - Early bubble						
Bubble factor $\left(\alpha_{blood} D_{blood}\right)$	0.50	0.25	0.47	0.57	0.86	0.28
Relative bubble factor ($N_2 = 1$)	1.8	0.88	1.7	2.0	3.0	1.0
Phase 2. - Terminal bubble						
Bubble factor $\left(\alpha_{fat} D_{fat}\right)$	0.14	0.08	0.41	0.85	2.50	0.22
Relative bubble factor ($N_2 = 1$)	0.64	0.34	1.9	3.9	12.0	1.0
CASE 3 Bubbles originating intravascularly in adipose or muscle tissue and lodging in the pulmonary or systemic vasculature						
Bubble factor (Relative rank)	2	1	3	4	5	2
CASE 4 Gas pockets in connective tissue						
Bubble factor (Relative rank)	2	1	3	4	5	2

[a] These factors hold for any given supersaturation ratio; after Roth (1965).

senting peak bubble size or pressure for each gas. Table 5 represents such calculations for four different classes of bubbles: Case 1, bubbles forming within fat tissue; Case 2, bubbles forming with blood vessels inside fat; Case 3, bubbles forming within blood vessels in fat or muscle and passing out to the lungs; and Case 4, bubbles forming gas pockets in connective tissue, such as the joints, where bends pain originates.

The higher the relative bubble factor, the greater the degree of hazard. It can be seen that in the case of bubbles forming within fat tissue, helium should be only one-fourth as dangerous as nitrogen, and neon only one-sixth as dangerous. Bubbles in fat can cause the most severe symptoms of neurocirculatory collapse but the incidence of this syndrome is low. For the more frequent bubbles forming in joint tissue or in the bloodstream, the calculations are less reliable but neon again appears safest, and helium equal to nitrogen in degree of danger. The gases argon, xenon, and krypton are much more dangerous than nitrogen in bubble sites. It must be emphasized that these calculations are only first approximations. One would expect the Case 1 condition to be independent of decay rate of inert gas concentration in the tissues and to be the most reliable. Navy diving experience suggests that the predictions for the Case 4 bubbles (bends symptoms) are also probably valid. Helium bubbles in the joints form as frequently, but more rapidly than, nitrogen bubbles in divers, and the symptoms are usually more short-lived. Helium appears to offer very little to the astronaut except possibly a reduction in the frequency and severity of the rare neurocirculatory collapse syndrome.

Neon has been used by the Royal Navy in diving with preliminary favorable results. Neon also has favorable characteristics for decreasing the lung damage in explosive decompression and in decreasing the hazard of ebullism, the formation of water vapor and gas bubbles at altitudes above 70,000 ft. These are less important considerations than classical decompression sickness.

Neon, therefore, looks like a favorable substitute for nitrogen in mixed-gas systems, but there are few specific data to support this conclusion. There is a dearth of knowledge about long-duration inhalation of $Ne-O_2$ mixtures. Moreover, little is known about the biochemical role of N_2 gas in the body and the effect of chronic exposure to atmospheres with no N_2. One would expect little trouble from the chronic lack of nitrogen or from the presence of helium or neon, but there are no clear-cut data available on this subject.

The role of inert gases in modifying radiation sensitivity of animals has also been inadequately studied. The work of Ebert and Howard (Table 6) suggests that very high pressures of helium, heon, and argon are required to alter the radiation response of plants

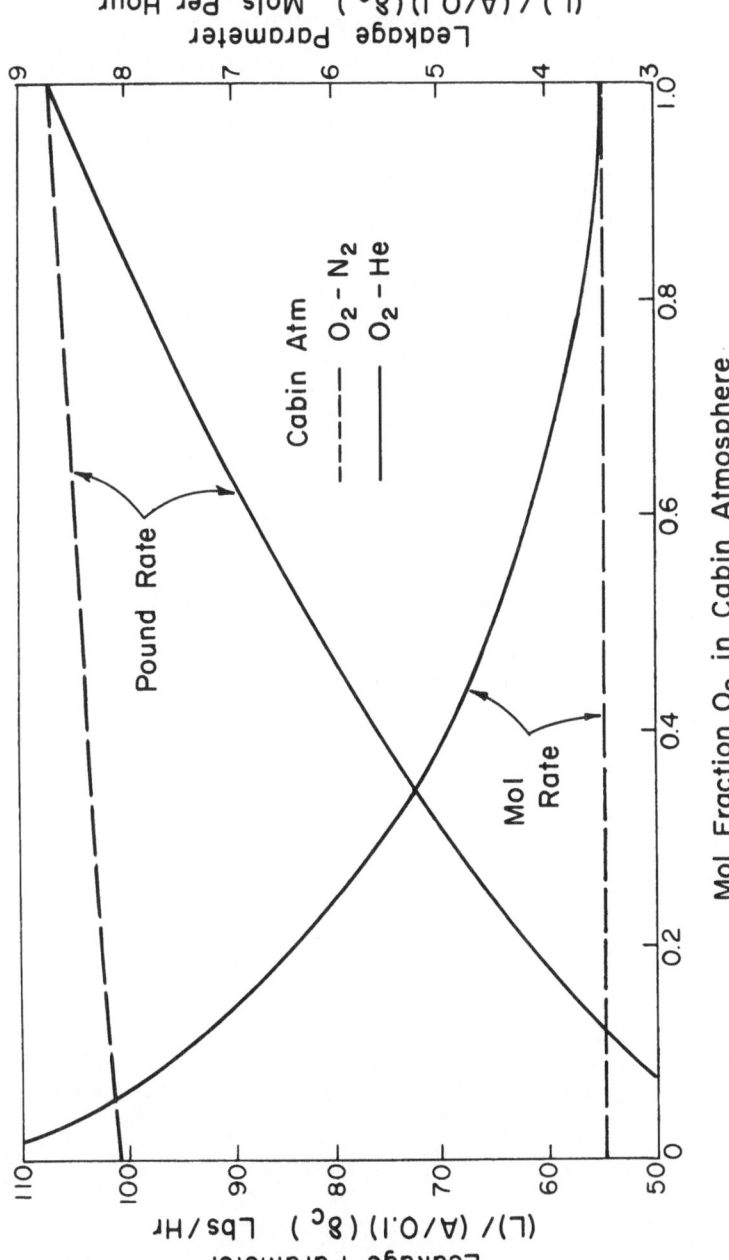

Figure 7. Variation in cabin leakage parameter with cabin oxygen concentration for oxygen–nitrogen and oxygen–helium atmosphere; after Dryden (1956).

TABLE 6

RADIOPROTECTIVE EFFECT OF METABOLICALLY INERT GASES[a]

Gas	Pressure (atm.) for 50% effect[b]
Helium	55
Hydrogen	55
Nitrogen	12.5
Argon	2
Xenon	1
Krypton	2

[a] After Ebert and Howard (1957).
[b] Pressure of metabolically inert gases added to 1 atm. of air to reduce the oxygen-dependent radiosensitivity of *Vicia faba* roots by 50%.

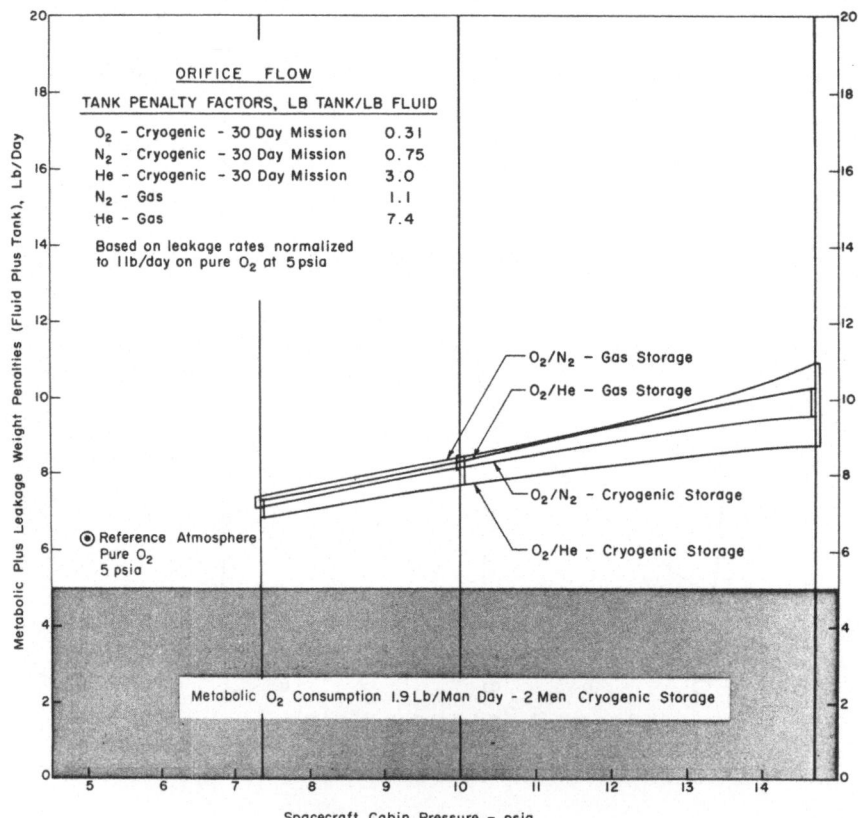

Figure 8. After Mason (1964).

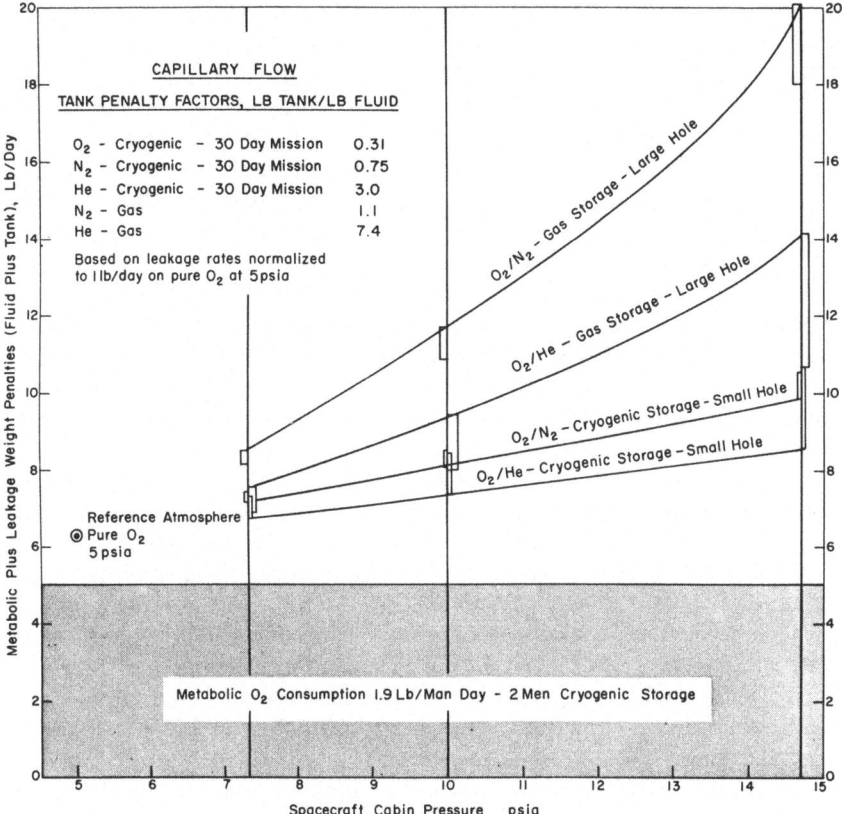

Figure 9. After Mason (1964).

(Ebert and Howard, 1957). One can predict that absence of inert gas or other alterations of our normal inert gas environment (608 mm. Hg nitrogen pressure) will probably have little effect on radiation sensitivity of mammals. Supporting data are certainly required on this point.

Finally, we come to the problem of leakage of gas mixtures from sealed cabins. In previous analyses, sonic-orifice equations have been used to show that helium mixtures will leak much more rapidly than nitrogen-oxygen mixtures. There is evidence, however, that capillary-diffusion leak conditions probably prevail (Roth, 1965; Mason, 1964). Figure 7 shows that the rate of leakage of a gas through a sonic-orifice must be viewed as both the molar and mass-rate of flow. In the most probable case of a 0.5 mole fraction, there is a slightly greater molar leakage of helium-oxygen, but about 1.3 times greater pound leakage of nitrogen-oxygen.

Mason (1964) has recently compared the capillary diffusion and sonic-orifice equations in the over-all weight penalty for 2-man-30-day missions with leak rate normalized to 1 lb./day on pure O_2 at 5 p.s.i. The tank penalty factors have been chosen from the best current tankage designs. It can be seen in Figure 8 for orifice flow there is little difference in weight penalty between 50% helium-50% oxygen and 50% nitrogen-50% oxygen mixtures for either gas-phase storage or cryogenic storage.

The calculations for capillary flow leakage are seen in Figure 9. There is little difference in over-all weight penalty for cryogenic storage of the mixtures. For gas storage with a hole 30μ in diameter (large hole), the nitrogen weight penalty is considerably greater at higher cabin pressures. Preliminary study of leakage factors suggests that capillary-diffusion flow is the mode of leakage through elastomer-metal seals (Mason, 1965).

Thus one can not rule out helium on the basis of leak penalty. The cryogenic storage of neon presents a lower tankage weight penalty than that of helium and the mass-rate of leakage should lie between that of helium and nitrogen. Therefore, neon may be more favorable than helium or nitrogen in over-all weight penalty.

We see, then, that even today selection of an ideal space cabin atmosphere represents a complex decision. It must be tailored to the mission and to our state of knowledge regarding many physical and biochemical variables. The more complex space station and interplanetary mission of the future further complicate the story. As in most scientific areas, however, early confusion usually leads to more complete understanding, simplification, and utilization.

REFERENCES

Dale, W. A., and H. Rahn (1952), Rate of gas absorption during atelectasis. Am. J. Physiol., 170:606.

Dryden, C. E., L-S. Han, F. A. Hitchcock, and R. Zimmerman (1956). Artificial cabin atmosphere systems for high altitude aircraft. TR-55-353, Wright-Patterson AFB, Ohio (November).

Ebert, M., and A. Howard (1957). Effect of nitrogen and hydrogen gas under pressure on the radiosensitivity of the broad bean root. Radiation Res., 7:331-341.

Kelton, A. A., and J. K. Kirby (1964). Total oxygen pressure and radiation mortality in mice. Douglas paper 2030 (May).

Klein, H. A. (1960). The effects of cabin atmospheres on bustion of some flammable aircraft materials. WADC TR-59-456 (ASTIA No. AD-238367).

Lewis, B., and G. von Elbe (1951). Combustion, flames and explosions of gases. Academic Press, New York.

Luft, U. C. (1962). Data on oxygen pressure effects compiled for the Garrett Corporation. The Lovelace Foundation, Albuquerque, New Mexico.

Mason, J. (1964). The two-gas spacecraft cabin atmosphere, engineering considerations. Internal Document SS-3223, AiResearch Manufacturing Co., Los Angeles, California.

_____(1965). Personal communication.

Roth, E. M. (1964a). Carbon dioxide effects. NASA-SP-3006, p. 8.

_____(1964b). Space-cabin atmospheres, P. I: Oxygen toxicity. NASA-SP-47.

_____(1964c). Space-cabin atmospheres, P. II: Fire and blast hazards. NASA-SP-48.

_____(1965). Space-cabin atmospheres, P. III: Physiological factors of inert gases. Lovelace Foundation, Albuquerque, New Mexico.

Welch, B. E., T. E. Morgan, Jr., and H. G. Clamann (1963). Time concentration effects in relation to oxygen toxicity in man. Federation Proc., 22:1053-1056.

3.

WEIGHT OPTIMIZATION OF FLIGHT TYPE CRYOGENIC TANKAGE SYSTEMS

BLASE J. SOLLAMI

The Bendix Corporation, Davenport, Iowa

INTRODUCTION

The storage of oxygen and hydrogen for space and lunar missions represents significant advances in the use of materials, fabrication techniques, vacuum technology, and thermal protection in conditions of varying gravity and thermal environments.

This study deals with the cryogenic storage supply systems of oxygen and hydrogen both for fuel cell operation and for life support in manned spacecraft and during lunar missions after extended periods of storage. For the large quantities of cryogen required, weight optimization of reliable storage systems is a prime prerequisite. The Dewar design concept quite readily satisfies the requirements for long-term storage of these cryogens in lightweight and reliable systems.

In essence, the basic requirement for all cryogenic storage systems is to minimize the heat that is transferred to the cryogen. The heat transferred in a Dewar-type storage system consists mainly of conductive and radiative heat transfer, as the convective heat transfer between the vacuum jacket and the inner vessel is essentially eliminated by maintaining a vacuum of less than 10^{-4} mm. mercury. Conductive heat transfer is minimized by using materials with low conductivity and also by increasing the length of path. Radiation heat transfer, on the other hand, is determined by the emissivity of the surfaces and the number of barriers separating the surfaces. Discrete, isothermally mounted shields fulfill the barrier requirement, whereas superinsulation (consisting of alternate layers of aluminum foil and Dexiglas paper) serves as a means of introducing many barriers, but increases the conductive heat transfer.

The purpose of this study was to provide information from which weight- and size-optimized hydrogen and oxygen storage and supply

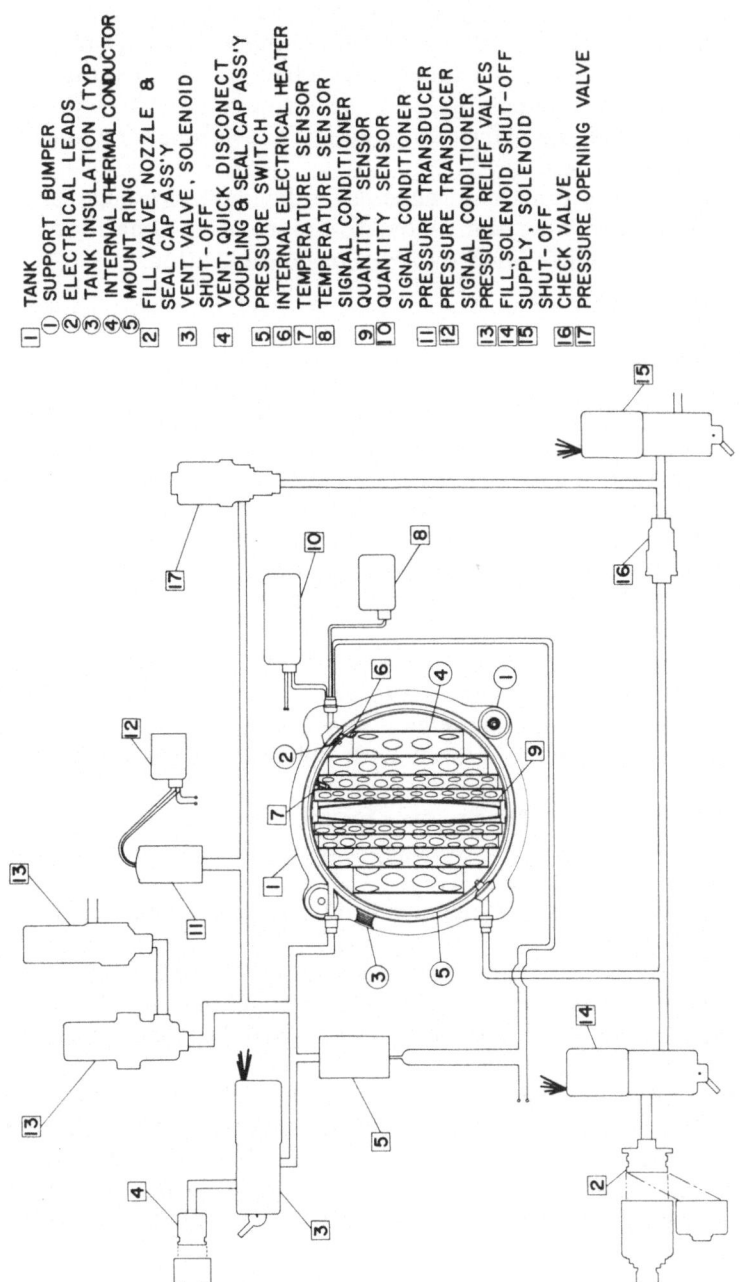

Figure 1. A simple schematic of a conventional spherical subcritical tankage system.

1 TANK
① SUPPORT BUMPER
② ELECTRICAL LEADS
③ TANK INSULATION (TYP)
④ INTERNAL THERMAL CONDUCTOR
⑤ MOUNT RING
2 FILL VALVE, NOZZLE &
   SEAL CAP ASS'Y
3 VENT VALVE, SOLENOID
   SHUT-OFF
4 VENT, QUICK DISCONECT
   COUPLING & SEAL CAP ASS'Y
5 PRESSURE SWITCH
6 INTERNAL ELECTRICAL HEATER
7 TEMPERATURE SENSOR
8 TEMPERATURE SENSOR
9 SIGNAL CONDITIONER
   QUANTITY SENSOR
10 QUANTITY SENSOR
   SIGNAL CONDITIONER
11 PRESSURE TRANSDUCER
12 PRESSURE TRANSDUCER
   SIGNAL CONDITIONER
13 PRESSURE RELIEF VALVES
14 FILL.SOLENOID SHUT-OFF
15 SUPPLY, SOLENOID
   SHUT-OFF
16 CHECK VALVE
17 PRESSURE OPENING VALVE

systems could be determined using the superinsulation concept. The approach followed in the investigation was to incorporate the major variables affecting system weight and a number of simplifying assumptions into a relatively simple computer program.

Figure 1 represents a functional schematic diagram of a subcritical storage and supply system. This figure identifies the functional components and shows their relative locations for a *spherical* storage configuration.

Figure 2 is a functional schematic diagram of a supercritical storage and supply system showing the relative locations of the functional components for a *cylindrical* configuration.

GUIDELINES

In the investigation of the long-term storage of liquid oxygen and liquid hydrogen, parametric data relating system weight and size for various quantities of cryogens are required. The system restraints and requirements that have a significant effect on this analysis are: quantity of cryogen; vessel shape; standby time; and minimum tankage pressure during operation. The following are the ranges of the variables considered for this analysis.

1. Quantity of cryogens (quantity of usable fluid for the standby time considered): Hydrogen, 15 to 120 lb.; oxygen, 200 to 1200 lb.

2. Vessel shape: Spherical vessels for oxygen storage; spherical or cylindrical vessel with spherical ends for hydrogen storage.

3. Standby time: Three, six, or nine months, including prelaunch, launch, and lunar storage; that is, the time elapsed before withdrawal of fluid for use.

4. Minimum tankage pressure during operation: 100 psia.

Two storage techniques have been investigated in this analysis. One is the nonvented storage concept, in which heat is absorbed at essentially a constant liquid bulk density until the relief valve pressure setting is reached at maximum standby time.

The vented concept extends the first technique in that, after reaching the pressure setting, additional heat input results in the vaporization and venting of fluid in subcritical systems. In supercritical units, the heat input contributes to fluid expansion and venting at the relief valve pressure, and is a function of the bulk density of the fluid.

The cryogenic storage systems considered in this investigation are complete in every detail as to the internal and external hard-

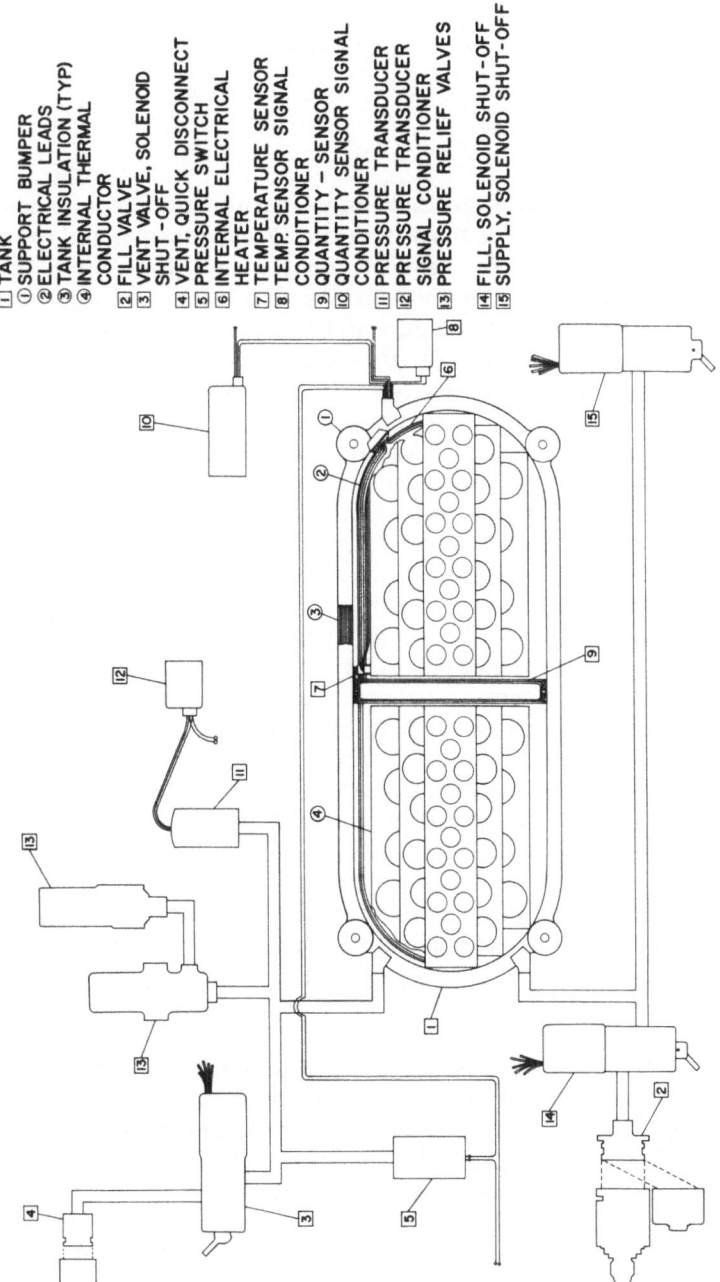

1. TANK
2. SUPPORT BUMPER
3. ELECTRICAL LEADS
4. TANK INSULATION (TYP)
5. INTERNAL THERMAL CONDUCTOR
6. FILL VALVE
7. VENT VALVE, SOLENOID SHUT-OFF
8. VENT, QUICK DISCONNECT
9. PRESSURE SWITCH
10. INTERNAL ELECTRICAL HEATER
11. TEMPERATURE SENSOR
12. TEMP. SENSOR SIGNAL CONDITIONER
13. QUANTITY - SENSOR
14. QUANTITY SENSOR SIGNAL CONDITIONER
15. PRESSURE TRANSDUCER
16. PRESSURE TRANSDUCER SIGNAL CONDITIONER
17. PRESSURE RELIEF VALVES
18. FILL, SOLENOID SHUT-OFF
19. SUPPLY, SOLENOID SHUT-OFF

Figure 2. A schematic diagram of a supercritical cylindrical storage and supply system.

ware, the sensors used to measure temperature, pressure and quantity of fluids, and the control and operating fixtures.

ANALYSIS

Subcritical storage

A subcritical fluid is one that has an equilibrium vapor pressure below the critical pressure of the fluid. For hydrogen and oxygen, the critical pressures are, respectively, 187 and 736 p.s.i.a. Under subcritical conditions, the fluid exists in a liquid and a gas phase which are easily distinguishable and separable. Cryogenic fluids are usually stored, shipped, and used in the subcritical state. Subcritical or two-phase cryogenic systems have been used to furnish breathing oxygen in nearly all military aircraft for over 15 years. A simple schematic of a conventional subcritical storage and supply system is shown in Figure 1.

In gravitational environments, an interface exists between the liquid and gas phases of a subcritical system, and the lighter gas phase is always oriented above the more dense liquid phase in the direction of the gravitational field.

The usage path of a subcritical cryogenic storage supply system is indicated on the temperature-entropy diagram in Figure 3. The line ABE denotes a constant-pressure subcritical supply system in which either liquid, vapor, or both may flow. The filled subcritical vessel is at operating pressure at point A. As fluid is withdrawn, the quantity of liquid decreases and the thermodynamic properties of the contents of the vessel are represented by the line AB. When all the liquid is removed, the line BE crosses the dome or two-phase region and only the gas phase remains. The line BE represents the properties of the gas phase. At E the contents remaining in the vessel are at the final temperature and pressure of the system.

Supercritical storage

A supercritical fluid is one that has an equilibrium vapor pressure above the critical pressure. In a supercritical state the fluid is in a single phase; hence, orientation or separation does not present a problem. This characteristic is extremely favorable when usage is required in a zero g environment.

A schematic of a typical cryogenic supercritical storage and supply system is pictured in Figure 2. The storage of fluids at cryogenic temperatures and supercritical pressures has been in various stages of development, undergone testing, and has been used

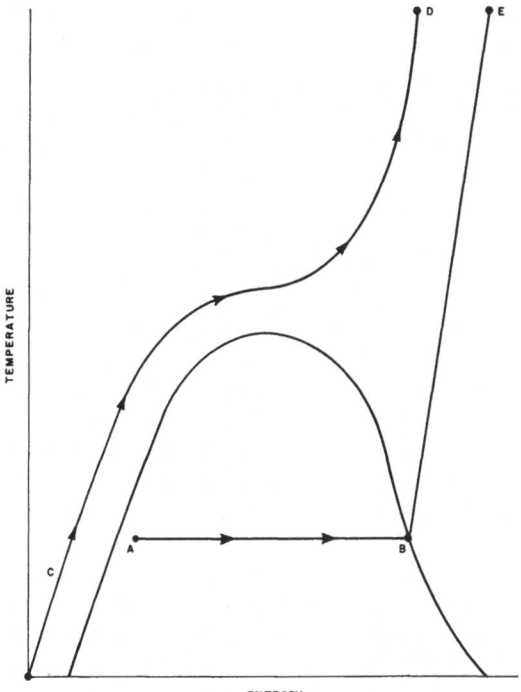

Figure 3. Temperature-entropy diagram
of the usage path of a subcritical cryogenic
storage supply system (*ABE*) and of a supercritical system (CD).

in applications for approximately 5 years.

The normal usage path of a supercritical cryogenic storage
and supply system is shown in Figure 3. The line CDE represents
the thermodynamic properties of the fluid at all stages of use. The
line segment CD represents depletion at constant pressure. At D
the pressure is allowed to decrease to the same final pressure and
temperature as the subcritical storage system.

Nonventing storage

The conventional nonventing cryogenic storage system allows
no vapor to be released from the vessel through the relief valve.
In essence, no fluid is lost from the system from the time the system is sealed, after being filled to a specified level, until the
vessel's contents have been used. The usable quantity of fluid in the
tank is therefore equivalent to the fluid contents minus the residual
remaining at the temperature and pressure of the system when the

prescribed mission is complete. The standby time between sealing and use is the total change in internal energy of the fluid divided by the rate of heat leak into the vessel.

A nonventing standby provides the simplest and most reliable cryogenic storage and supply system. In general, a nonventing storage system is weight economical for cryogens, such as oxygen, that require a low system weight per pound of fluid stored.

Venting storage

The venting storage system differs essentially in procedure. In a manner similar to that used for the nonventing system, the vessel is filled with the required quantity of fluid and is sealed off. A predetermined quantity of the fluid is allocated for venting. The standby time consists of an initial nonventing period during which the pressure rises from atmospheric to the venting relief valve pressure and, hence, of an additional period for venting the allocated quantity of fluid.

In addition to the heat leak into the vessel required to build up the pressure to the relief valve pressure setting, the vented fluid also requires a significant quantity of heat. The heat required for venting fluid is dependent on the mode of storage. In subcritical storage systems a fixed amount of heat input is necessary for each pound of fluid vented. On the other hand, for supercritical storage systems the heat input required to maintain pressure for each pound vented is a function of the density of the fluid.

Long venting standby periods impose a large number of cycles on the venting flow control regulator (or venting relief valve), therefore it is necessary to use matched, series relief regulators to obtain a reasonable reliability. A venting standby is weight economical over a nonventing standby for fluids with a high system-weight to fluid-weight ratio, such as hydrogen.

Parameters and assumptions

In general, the cryogenic storage system weight is most affected by the parameters subsequently described for this study. The range of these parameters was determined by the current state of the art. In some instances the range was governed by a realistic foreseeable improvement in the state of the art in materials and manufacturing techniques.

The cryogenic storage systems considered here are presented in complete detail regarding both internal hardware for measurement of the pressure, temperature, quantity of fluid in the vessel, thermal conductor, and external hardware for operation and pressure control of the system.

Recent developments in materials have resulted in alloys that possess a high strength to weight ratio. Inconel 718 and Titanium 6A1-4V have been shown to be compatible with hydrogen, and Inconel 718 with oxygen. Both of these materials possess good structural properties at the cryogenic temperatures. These materials were used in this study for the inner vessel, connecting tubing, vacuum jacket, and vessel fittings for the respective cryogen. State-of-the-art fabrication techniques currently exist for both these alloys.

The external plumbing and components for the control and operation of both cryogenic storage vessels are essentially the same. The weight breakdown shown in Table 1 is based on currently available components, which include a reduction in weight for an improvement in the state-of-the-art in design and weight optimization.

TABLE 1

EXTERNAL PLUMBING

Component	Weight (lb.)
Fill valve (solenoid)	0.50
Vent valve (solenoid)	0.50
Relief valve	0.471
Shut-off valve (solenoid)	0.50
Check valve	0.140
Pressure switch	0.20
Pressure transducer	0.037
Pressure signal conditioner	0.333
Quantity signal conditioner	0.707
Boss	0.20
Getter and blowout patch	0.180
Total	3.768

An example weight breakdown for the internal components and mounting structures, which are a function of the vessel size, is shown in Table 2. These data are based on a vessel having an internal diameter of 43.5 in.

The minimization of system weight for the cryogenic storage concepts considered is markedly affected by that hardware which is dependent upon either the size or weight (or both) of the vessel and the operational requirements imposed on the system. As a con-

TABLE 2

INTERNAL PLUMBING AND TANK SUPPORTS

Item	Weight (lb.)
Weld doubler	1.50
Bumpers (inner vessel support)	3.20
Tubing (between inner and outer vessel)	0.40
Fittings	0.25
Electric heater	0.25
Internal thermal conductor	8.00
Quantity sensor	1.35
Temperature sensor (PRT)	0.25
Mount ring (internal component)	1.82
Miscellaneous external hardware	1.00
Mounting carriage assembly	35.00
Total	53.02[a]

[a]This sum is for an oxygen system.

sequence, these components and structures lend themselves quite readily to weight optimization as some function of system weight or volume.

The total weight given in Table 2 is for an oxygen system. An equivalent size hydrogen system has a mounting carriage assembly weight estimated at 20 lb.

Since the density of the oxygen is high, the weight of the internal plumbing and tankage support systems was assumed to be directly proportional to the external diameter of the storage vessel. For the hydrogen storage system, this weight is related to the volume of the inner vessel. The dependency of this equipment on the size and quantity of fluid involved requires further refinement for the large vessels considered in this investigation.

The heat input to the cryogenic fluid through the tank supports or bumpers was based on recently gathered thermal conductivity data on the proposed bumper systems. The actual data have been reduced slightly to account for a reasonable expected improvement in bumper design. Table 3 is an estimate of the hourly heat input through the bumper systems for various size vessels under various gravitational conditions. These data are for heat transfer through the bumper system for a temperature gradient from liquid oxygen temperature (160°R.) to room temperature (540°R.).

These data were fitted by an exponential-type curve with these resulting equations for a daily heat input:

Earth $Q = 14.4 + 15.8904 \exp(-0.126V)$

Orbit $Q = 2.88 + 3.1766 \exp(-0.126V)$

Lunar $Q = 8.64 + 9.5352 \exp(-0.126V)$

where V is the volume of the inner vessel in cubic feet.

TABLE 3

THERMAL HEAT TRANSFER THROUGH BUMPER SYSTEM

Storage condition	Vessel size (cu. ft.)	Heat input (B.t.u./hr.)
Earth	4.0	1.0
	15.0	0.7
	∞	0.6
Orbit	4.0	0.20
	15.0	0.14
	∞	0.12
Lunar	4.0	0.6
	15.0	0.42
	∞	0.36

Several concepts for the insulation of the cryogenic vessels should be considered. Among them are the simple vacuum jacket and a vacuum with either discrete radiation barrier shields or laminar superinsulation, or both (the aforementioned concepts incorporated vapor cooling). Additionally, the insulation can be packaged in a rigid evacuated annular space load bearing as with a collapsible outer vacuum shell or as sealed but nonevacuated insulation layers. For this study, the initial approach is to use only a rigid outer vacuum shell with evacuated superinsulation.

The effectiveness of Linde S1-44 superinsulation for both flat and cylindrical surfaces has been established. However, the application of superinsulation to compound surfaces such as spheres markedly decreases the effectiveness of the superinsulation, and the deviation increases with decreasing spherical radius. This effect has been attributed to the short-circuiting of the alternate radiation shields due to the extreme difficulty in applying the super-insulation to a compound curvature surface. Thus, it is assumed that the achievable thermal conductivity for the superinsulation used in the cryogenic storage vessels is double the reported value of 0.00002 B.t.u./(hr.) (sq. ft.) (°F./ft.). Furthermore, it is assumed that this value would prevail for both vessels independent of temperature.

The weight minimization is achieved by a weight trade-off between thickness of insulation and structural weight of the inner and outer vessels for both nonvented and vented storage concepts. The density of Linde S1-44 insulation at 70 layers/in. is 4.7 lb./cu. ft., or 0.00272 lb./cu. in.

The heat input through the optimum density of evacuated super-insulation (70 layers/in.) for concentric spherical insulation is given by:

$$Q = \frac{\pi k \, D_1 D_2 (T_2 - T_1)}{6(D_2 - D_1)}$$

where Q = total heat input (B.t.u./day), k = thermal conductivity of superinsulation Linde S1-44 = 0.00048 B.t.u./(day)(sq. ft.)(°F./ft.), D_1 = inside diameter of insulation or outside diameter of inner vessel (inches), D_2 = outside diameter of insulation or inside diameter of vacuum jacket (inches), T_1 = temperature of the cryogenic fluid (°R.), and T_2 = temperature of the outer surface of the vessel (°R.).

Similarly, for a cylindrical section the daily heat transferred is given by:

$$Q = \frac{\pi k (T_2 - T_1) \text{ (length)}}{6 \ln(D_2/D_1)}$$

Hence, the total heat input to a particular vessel is the sum of that transferred through the superinsulation and the bumper system. The resistance to heat transfer of the metal vessel is considered negligible and is not considered.

It is assumed that the cryogenic fluid is in equilibrium at its normal boiling point at atmospheric pressure when the vent valve is closed. A simplifying assumption is made that the contents of the vessel remain at this temperature throughout the pressurization and venting period. This simplification results in a conservative heat leak into the contents of the cryogenic storage vessel, but the effect on the general trends for weight optimization is negligible.

The initial percentage fill as used here is of fluid in the vessel to total weight of fluid the vessel will hold. In subcritical storage concepts this initial percentage fill was based upon the vessel being filled with liquid in equilibrium at the temperature and pressure prior to venting. In the venting concept for subcritical storage, the ullage at the onset of venting was considered to be zero. Tests have indicated that the quantity of fluid lost during this period is less than if an ullage existed.

The shell thickness for the spherical sections was determined on the basis of a safety factor of 2 based on the ultimate strength.

Inner pressure vessel

The thickness of the inner spherical pressure vessel was determined by:

$$t' = \frac{(P)(SF)(D)}{(4)(UTS)}$$

where $t = [t' + (t')^2/D]$; P = venting pressure (p.s.i.a.); SF = safety factor; D = inside diameter (inches); UTS = ultimate tensile strength (p.s.i.); t' = intermediate thickness based upon D (inches); t = shell thickness based upon average D (inches). For the cylindrical sections, the shell thickness is assumed to be double that for the spherical sections.

Vacuum jacket

The maximum pressure differential experienced by the outer vessel or vacuum jacket is 1 atm. Since this vessel has an external pressure load, the following buckling equation for spherical shells is used to calculate the thickness of the vacuum jacket. This equation is based on empirical theory with a correction factor included.

$$t = D_2 \sqrt{\frac{(P_1)(SF)\,3(1 - \nu^2)}{8(E)(C)}}$$

where D_2 = inside diameter of the vacuum jacket (inches); P_1 = external pressure (14.7 p.s.i.a.); ν = Poisson's ratio; E = modulus of elasticity (p.s.i.); C = correction factor (0.5).

The correction factor C was determined from actual buckling test data on hemispheres made of aluminum and stainless steel of varying wall thickness. The thickness of the outer vessel cylindrical sections was determined by the equation for buckling of a cylindrical shell under axial and uniform lateral pressures, which is

$$(SF)(P_1)$$

$$= \frac{2(E)(t)}{D_2(n^2 + \frac{1}{2}(\pi D_2/2L)^2)} \left[\frac{1}{(n^2(2L/\pi D_2)^2 + 1)^2} + \frac{(t^2)(n^2 - (\pi D_2/2L)^2)^2}{3D_2^2(1 - \nu^2)} \right]$$

where n = number of circumferential lobes into which the cylinder will buckle, and L = cylindrical length (inches). The physical prop-

TABLE 4

PHYSICAL PROPERTIES OF STRUCTURAL METALS

Property	Inconel 718	Titanium 6A1-4V
Ultimate tensile strength (p.s.i.)	196,000	130,000
Modulus of elasticity (p.s.i.)	29.6×10^6	16.5×10^6
Poisson's ratio	0.29	0.327
Density (lb./cu.in.)	0.297	0.161

erties of the metals considered are listed in Table 4.

The total storage time includes three distinct periods: The first is that prior to launch (estimated at 30 hr.); the second includes that time from earth launch to lunar landing, which will be called the orbiting period (estimated at 3 days); the final period is that time from lunar landing until operation begins. The sum of these three periods is included in the total days of lunar standby.

Under normal operational procedures, the cryogenic storage tanks will attain an equilibrium temperature dependent on their location in or on the space vehicle, and on the temperatures of the various components of the vehicle as well as the earth, orbiting, and lunar environment. During the earth stay, it is assumed that the ambient temperature is 160°F. While on the lunar surface, the over-all daily average temperature of the lunar surface is assumed to be −54°F. This latter average temperature was based on a 28-day lunar month having an average temperature of 175°F. for 4 days, −15°F. for 14 days, and −200°F. for 10 days.

It is further assumed that the outer surface temperature of the cryogenic storage vessels is the same as the lunar surface or the ambient and surrounding temperatures. This simplification will naturally result in a conservative calculation for the heat input to the cryogenic fluids. Since the lunar surface is believed to be covered with dust, it was considered probable that the cryogenic storage tanks would also be covered with this material. If such be the case, for lunar dust having an absorptivity, α, and an emissivity, ϵ, the resulting $\alpha{:}\epsilon$ ratio of 0.93 would exist, which is very close to the value of 1, which is assumed.

The degree of thermal stratification was neglected for both subcritical and supercritical storage systems. However, test results on a liquid oxygen storage vessel equipped with an internal thermal conductor indicate that an 80 to 85% approach to thermal equilibrium is achievable. Further testing and improved thermal conductor design concepts are necessary, both to verify and to improve these values.

At the conclusion of a mission, a quantity of fluid will remain in the storage vessels. The amount of fluid remaining will depend on the power requirement available to raise the temperature of the vapor, the tank pressure, and the size of the tank. For this investigation, the final pressure in the cryogenic systems is 103 p.s.i.a. (7 atm.). The resulting bulk densities of the residual vapor for oxygen and hydrogen are assumed to be 0.9 and 0.35 lb./cu. ft.

The physical and thermodynamic properties of parahydrogen and the properties of oxygen were taken from the literature (Roder et al., 1963; Stewart et al., 1963). The heat required to pressurize a vessel containing a cryogenic fluid to a subcritical pressure is in general given by

$$Q_P = W_{g,2}E_{g,2} + W_{L,2}E_{L,2} - W_{g,1}E_{g,1} - W_{L,1}E_{L,1}$$

where Q_P = total heat input required (B.t.u.); W = weight of fluid (lb.); E = internal energy of the fluid (B.t.u./lb.); g = gas; L = liquid; 1 = initial state; and 2 = final state.

The heat required to pressurize a vessel containing a cryogenic fluid to a supercritical pressure is dependent on the percentage fill and is given by

$$Q_P = H_2 - \frac{1}{\rho_2}\left[\frac{F - \rho_{g,1}/\rho_{L,1}}{1 - \rho_{g,1}/\rho_{L,1}}H_{L;1}\rho_{L,1}\right.$$
$$\left. + \frac{1-F}{1-\rho_{g,1}/\rho_{L,1}}H_{g,1}\rho_{g,1} + C_3(P_2 - P_{L,1})\right]$$

where H = enthalpy of fluid (B.t.u./lb.); ρ = density of fluid (lb./ cu. ft.); F = fraction fill; C_3 = conversion constant (0.18505); and P = pressure (p.s.i.a.).

Heat required to maintain pressure

In subcritical storage, the heat input required to maintain pressure while venting gaseous vapor from the gas phase after pressurization is the heat of vaporization of the fluid at the pressure in question. However, in supercritical storage the heat input is a function of fluid density. The heat input required to maintain pressure in a supercritical storage system per pound of fluid withdrawn is given by

$$Q/W = -\rho\left(\frac{dH}{d\rho}\right)_p$$

The variations of heat input for various densities of oxygen were plotted for 800 and 1,000 p.s.i.a. Since in the vented standby conditions the heat required to withdraw a pound of fluid increases the total heat input to the system, it is necessary to determine the total heat input during venting. Thus, equations for the heat input curves were developed; they are:

At 800 p.s.i.a.:
$$Q/W = \frac{1 - 0.048084664\rho + 0.0010798977\rho^2}{0.01466615 + 0.00004377986\rho}$$

At 1,000 p.s.i.a.: $Q/W = 0.059999989\rho^2 - 2.7666655\rho + 67.666645$

Similarly, the variations of heat input for various parahydrogen densities were plotted for pressures of 146.9, 220.4, 293.9, 367.4, and 661.3 p.s.i.a. The equations for some of these curves follow.

At 220.4 p.s.i.a.:

0.0 to 2.0 lb./cu. ft.:
$$Q/W = \frac{79400\rho^2 - 190853\rho + 284714}{664 + 814\rho}$$

2.0 to 4.5 lb./cu. ft.:
$$Q/W = \frac{127865\rho^2 - 433178\rho + 575504}{664 + 814\rho}$$

At 367.4 p.s.i.a.:

$$Q/W = \frac{3656\rho^2 - 13166\rho + 23928}{68 + 13\rho}$$

At 661.3 p.s.i.a.:

$$Q/W = \frac{0.15295327\rho^2 - 0.50876967\rho + 1}{0.0008675888 + 0.0010668968\rho}$$

Oxygen storage concept

Both vented and nonvented storage concepts for subcritical and supercritical pressures were investigated. The total standby time in days (lunar standby) was determined for varying thicknesses of insulation for nonvented and vented conditions. In the subcritical storage systems, the initial percentage fill was a direct result of the final pressure reached by the system. For supercritical storage, the initial percentage fill was varied, along with the variables of insulation thickness and final pressure, for nonvented and vented storage systems.

From the results of the computer analysis, graphs showing the effect of inner vessel pressure on total system weight and length of standby time for the nonvented and vented conditions in usable fluid range from 200 to 1,200 lb. were made; sample graphs are presented and discussed.

Figure 4 shows the subcritical spherical oxygen system weight as a function of standby time for 400 lb. of usable fluid. This graph is plotted for vented and nonvented conditions at pressures of 103, 220, 441, and 661 p.s.i.a. As shown, the nonvented systems are lighter than the vented systems for standby time of 180 days or less. It was not necessary to plot the curves for vented weights greater than 10 lb. as there is no indication in the computer analysis that a lighter system weight would be attained. Similar graphs were also plotted for weights of usable oxygen ranging from 200 to 1,200 lb. The final pressure of the system prior to venting does affect the total system weight, and as the standby time increases, the optimum subcritical pressure increases.

For the supercritical condition, it was not necessary to plot the vented condition as several vented condition points were checked, and they all resulted in greater system weights. Figure 5 shows the variation of system weight and standby with fill density for 800 p.s.i., supercritical system, and 400 lb. of usable fluid. The minimum weight system is attained at decreasing initial fill densities as the standby time increases. As in the subcritical condition, a smooth curve drawn tangent to the plotted curves in the area in which they are the minimum, results in a curve of minimum system weight versus standby for the intermediate fill densities. These graphs were drawn for from 200 to 1,000 lb. of usable oxygen at 800 and 1,000 p.s.i.a. internal pressures.

From these plots, the total system weight versus pressure was drawn for standby times of 90, 180, and 270 days for usable fluid of 200, 400, etc., to 1,200 lb. Figure 6 is an example for 400 lb. of usable oxygen that shows an optimum pressure which gives minimum system weight, and this pressure is subcritical over the entire range investigated.

Plots of weight of usable oxygen versus total system weight were then drawn from the foregoing data for 90-, 180-, and 270-day standby times and are shown in Figures 7, 8, and 9. Included are reference points showing the pressure and insulation thickness (T2) at several points; as the weight of usable oxygen increases, the pressure and insulation thickness decreases.

Hydrogen storage concept

In addition to the vented and nonvented concepts, both cylindrical and spherical storage systems were considered for hydrogen in

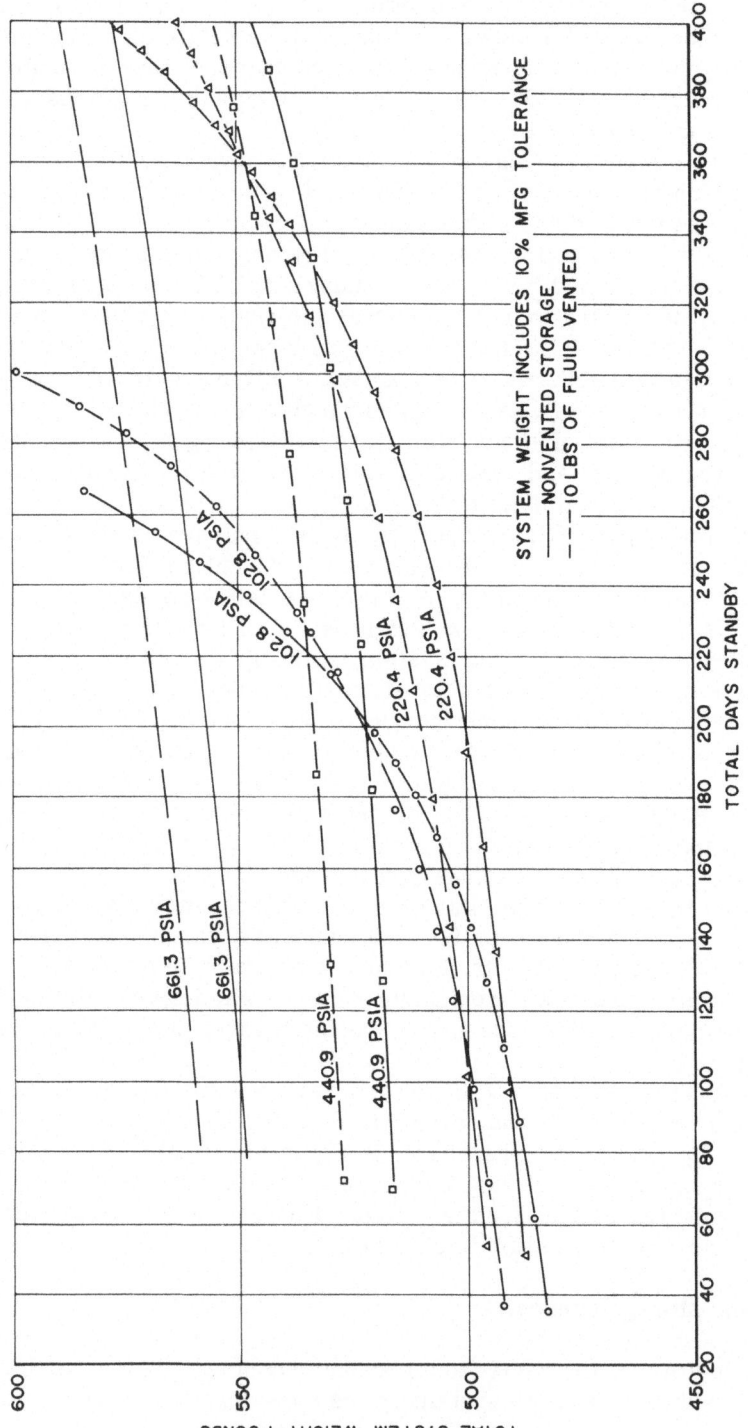

Figure 4. Subcritical spherical oxygen system weight versus standby time and pressure parameters for 400 lb. of usable fluid.

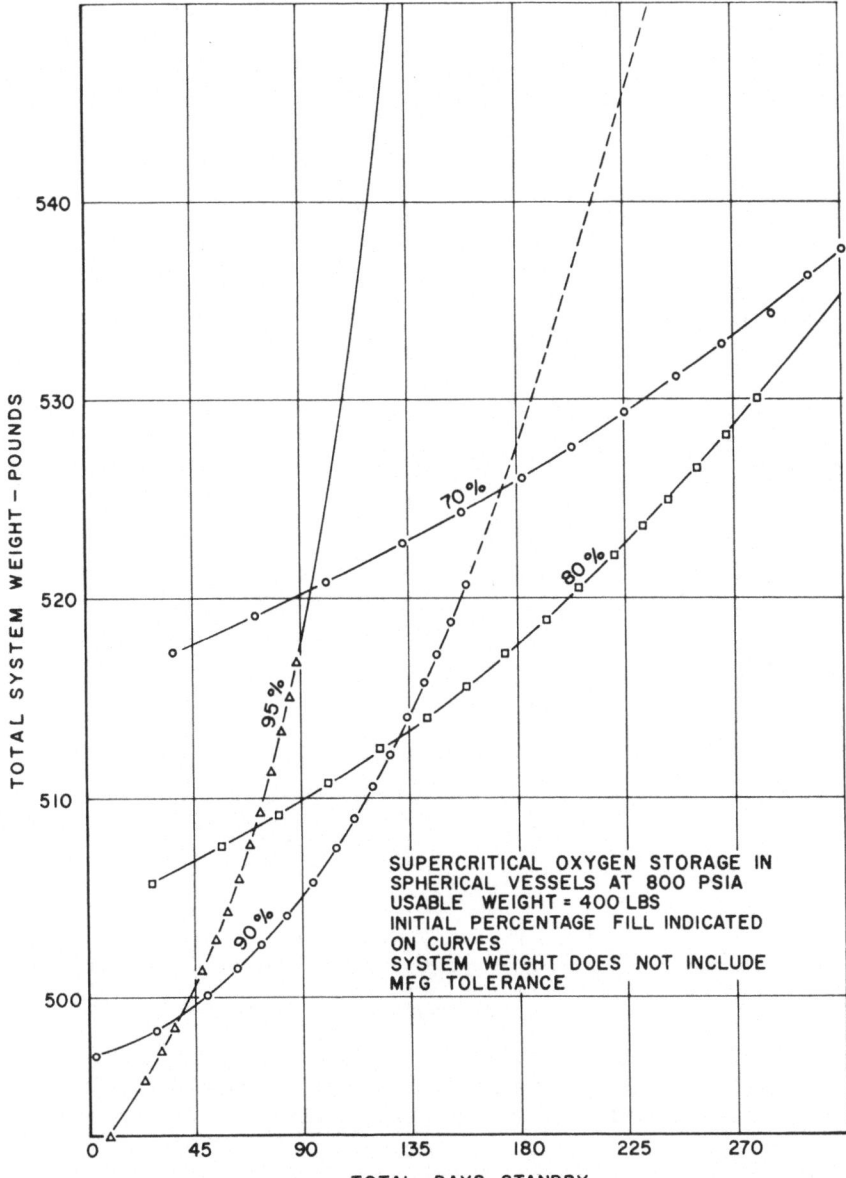

Figure 5. Supercritical oxygen system weight versus standby time and fill density for 800 p.s.i. and 400 lb. usable fluid. Initial percentage fill is indicated on curves; system weight does not include manufacturing tolerance.

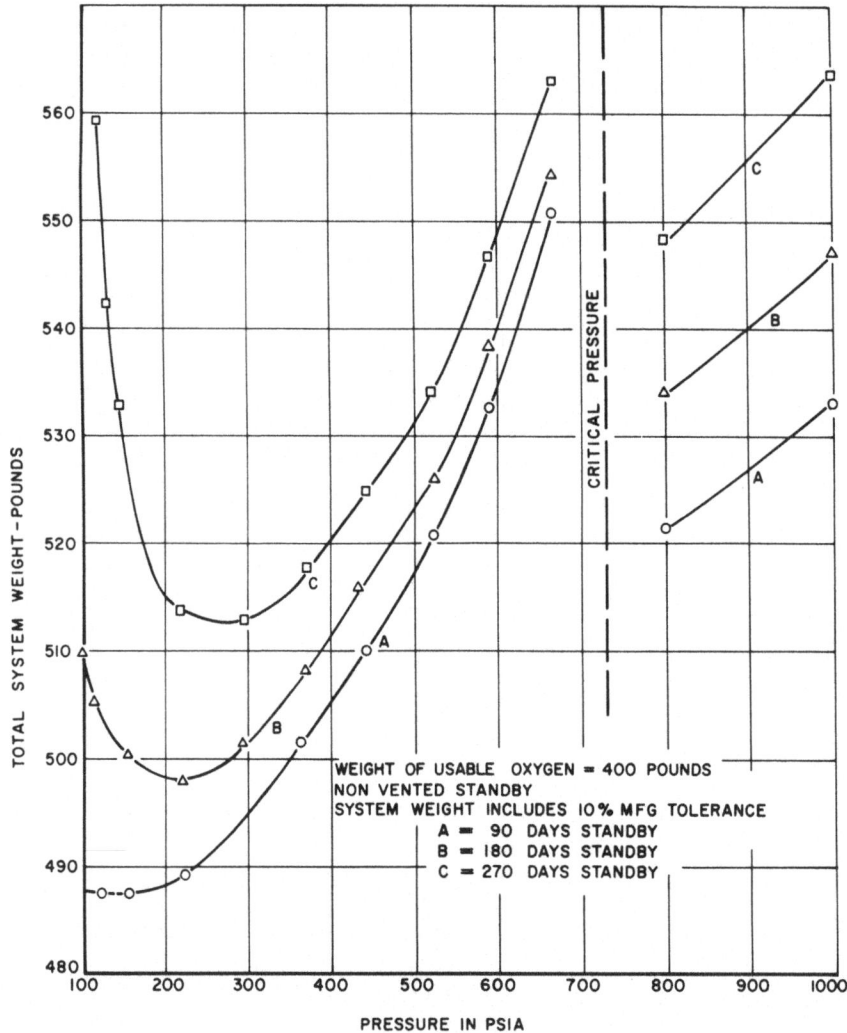

Figure 6. Minimum oxygen system weight versus pressure for 400 lb. of usable fluid. Nonvented standby; system weight includes 10% manufacturing tolerance; A, 90 days standby; B, 180 days standby; C, 270 days standby.

the subcritical and supercritical conditions. The cylindrical systems considered have L:D ratios of 2, 3, and 5. The procedure used in analyzing the computer data for the hydrogen system is essentially the same as that for the oxygen system; only representative plots are shown.

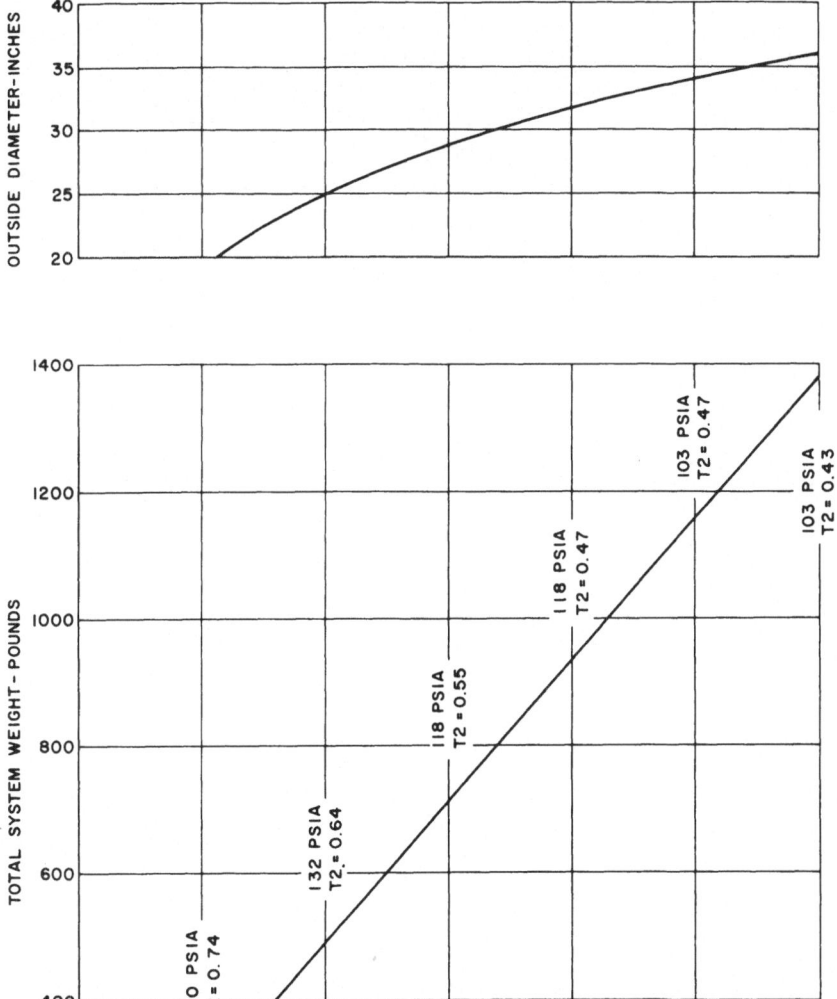

Figure 7. Optimum oxygen system weight and size versus usable fluid for 90 days standby. System weight includes 10% manufacturing tolerance; T2 is the insulation thickness in inches.

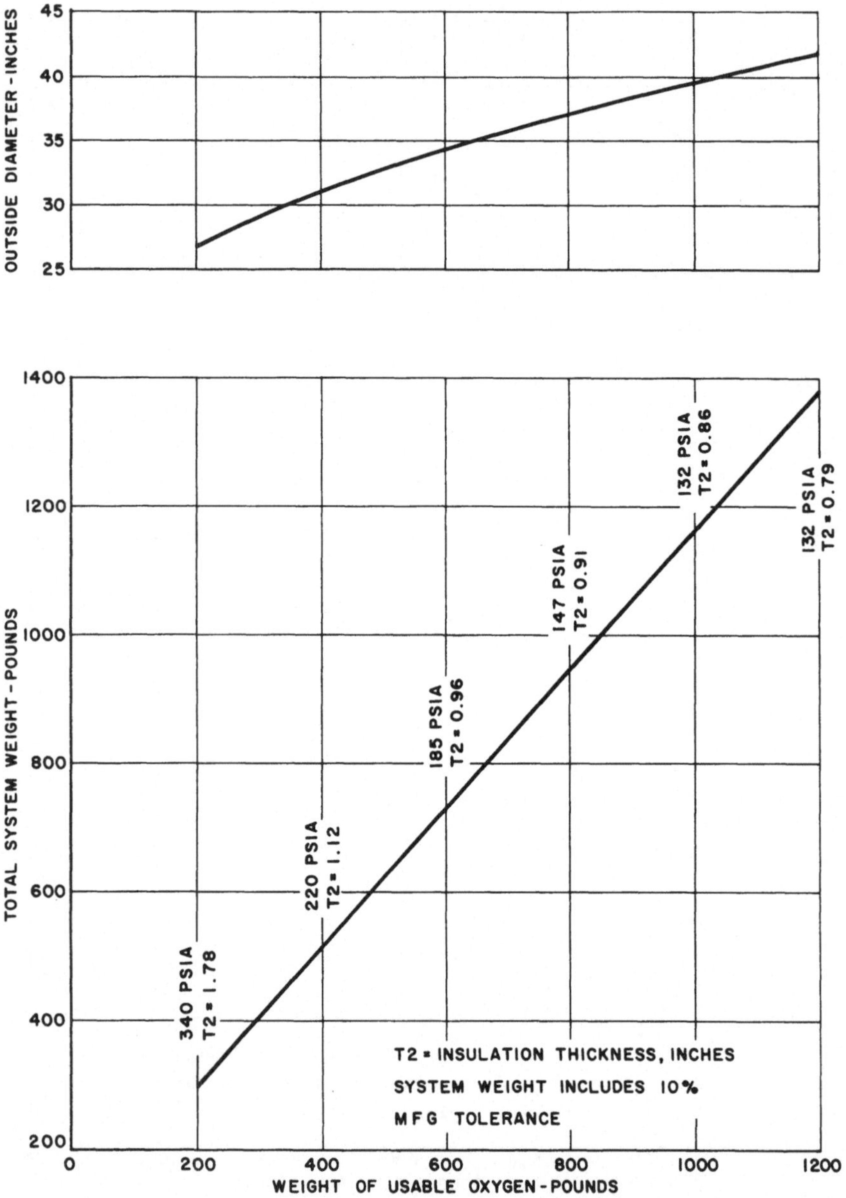

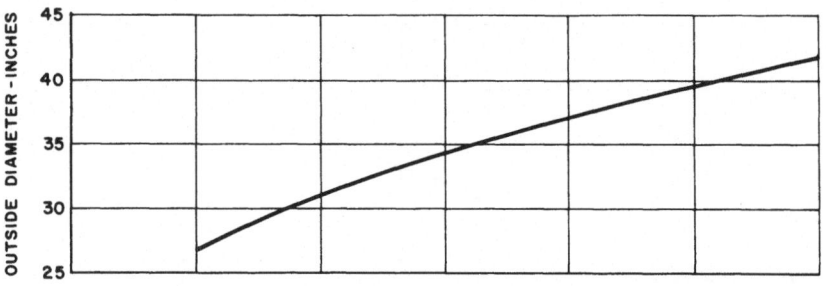

Figure 8. Optimum oxygen system weight and size versus usable fluid
for 180 days standby. System weight includes 10% manufacturing tolerance;
T2 is insulation thickness in inches.

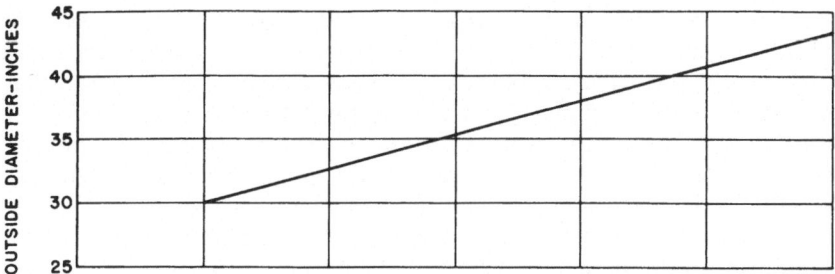

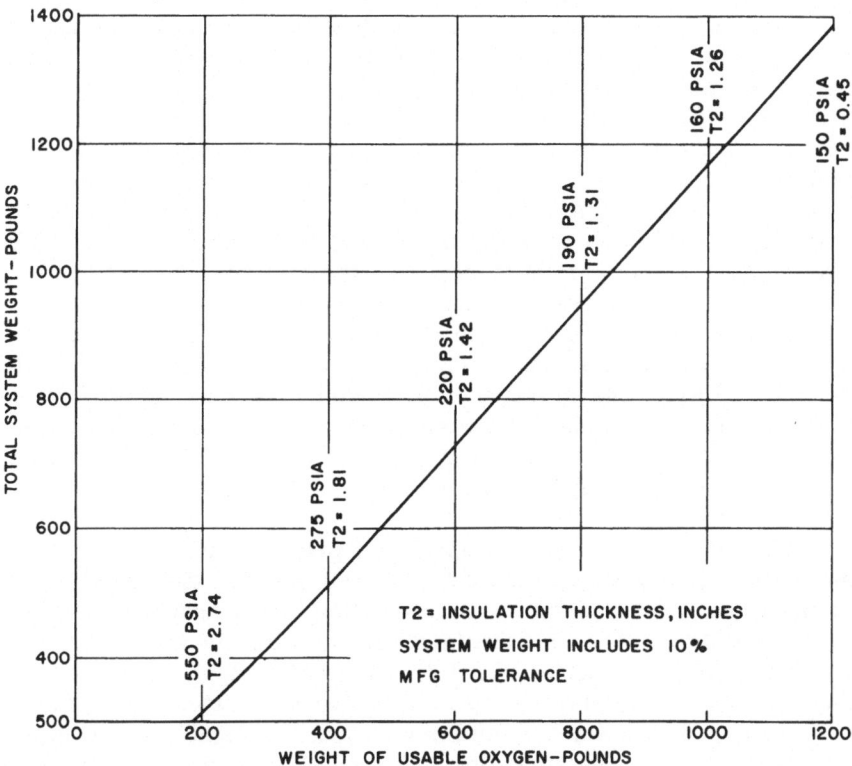

Figure 9. Optimum oxygen system weight and size versus usable fluid for 270 days standby. System weight includes 10% manufacturing tolerance; T2 is insulation thickness in inches.

For the subcritical hydrogen storage, plots of total system weight versus standby time at various subcritical pressures and usable fluid quantities were made. These graphs also indicate that for subcritical hydrogen storage, the lowest pressure system will result in a minimum weight. Figures 10 and 11 contain sample graphs for storage of 30 lb. of usable fluid in spherical and cylindrical (L:D = 3) vessels at a subcritical pressure of 103 p.s.i.a.

Similar plots were made for supercritical hydrogen storage. In addition to the pounds of vented hydrogen and thickness of insulation, the supercritical storage concept incorporates the percentage of initial fill in the analysis for the determination of minimum system weight. Figures 12 and 13 show system weight versus standby time at 661.32 p.s.i.a. for supercritical spherical and cylindrical (L:D = 3) vessels containing 30 lb. of usable fluid and 95% fill. Plots of these parameters at different fill densities and different usable fluid weight indicated that the minimum weight systems are those having an initial fill density of 95% for both cylindrical and spherical systems.

From the foregoing figures for hydrogen storage in both subcritical and supercritical concepts, the effect of pressure on total system weight for spherical and cylindrical vessels having a L:D = 3 is presented in Figures 14 and 15. These values are for 180 days of standby. It is evident that the minimum weight system occurs at storage pressures above the critical pressure.

The effect of pressure on the lunar storage of hydrogen for spherical vessels is presented in Figures 16 and 17 for 90 and 180 days. Similarly, the storage of hydrogen in a lunar environment for cylindrical vessels is presented in Figures 18 and 19. From these figures it is noted that for a given weight of usable hydrogen there is an optimum pressure which will result in a minimum weight system.

The effect of vessel shape at a pressure of 661 p.s.i.a. is shown in Figures 20, 21, and 22 for 90, 180, and 270 days of storage in the lunar environment considered. For this pressure, the minimum weight system is obtained for the storage of hydrogen in spherical vessels. Also, the total system weight increases as the aspect ratio (L:D) of the cylindrical storage systems increases.

Figure 23 shows the minimum system weight for a venting hydrogen storage system using cylindrical-shaped vessels having an aspect ratio of 3 as a function of the weight of usable hydrogen. The diameter of these vessels is also shown.

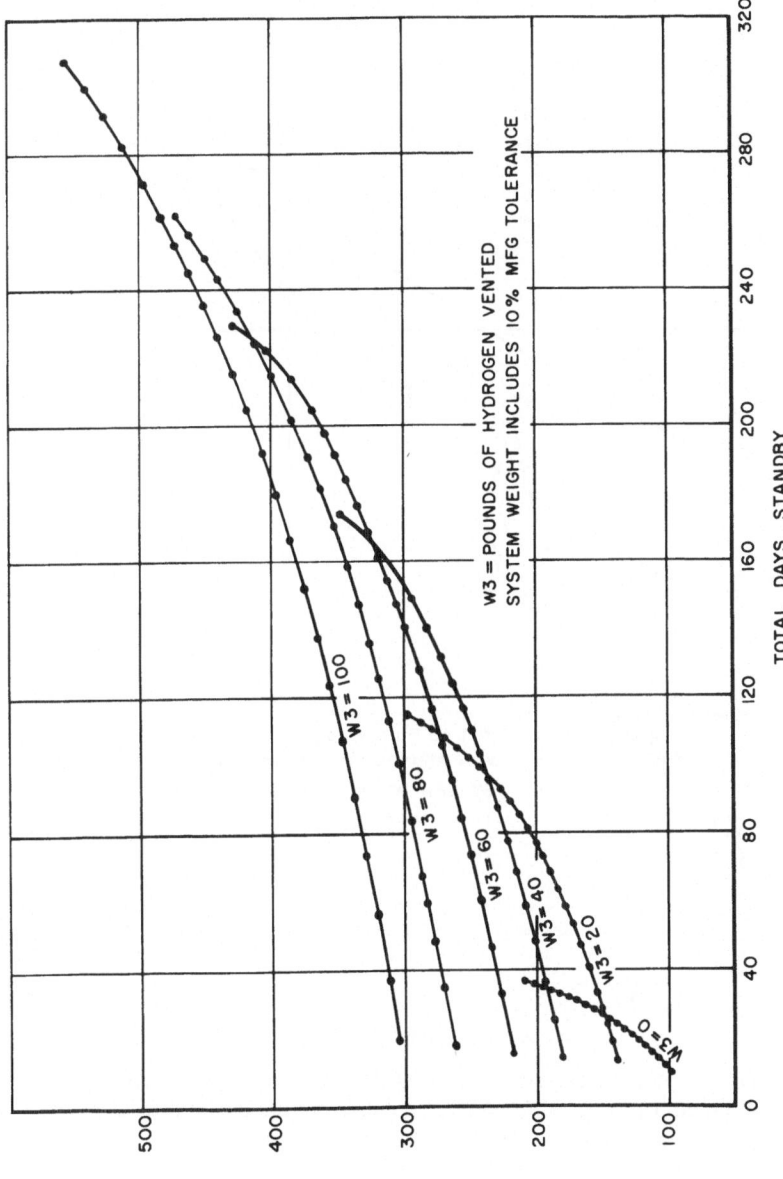

Figure 10. Subcritical spherical hydrogen system weight versus standby at 102.8 p.s.i.a. for 30 lb. of usable fluid. System weight includes 10% manufacturing tolerance; W3 is the number of pounds of hydrogen vented.

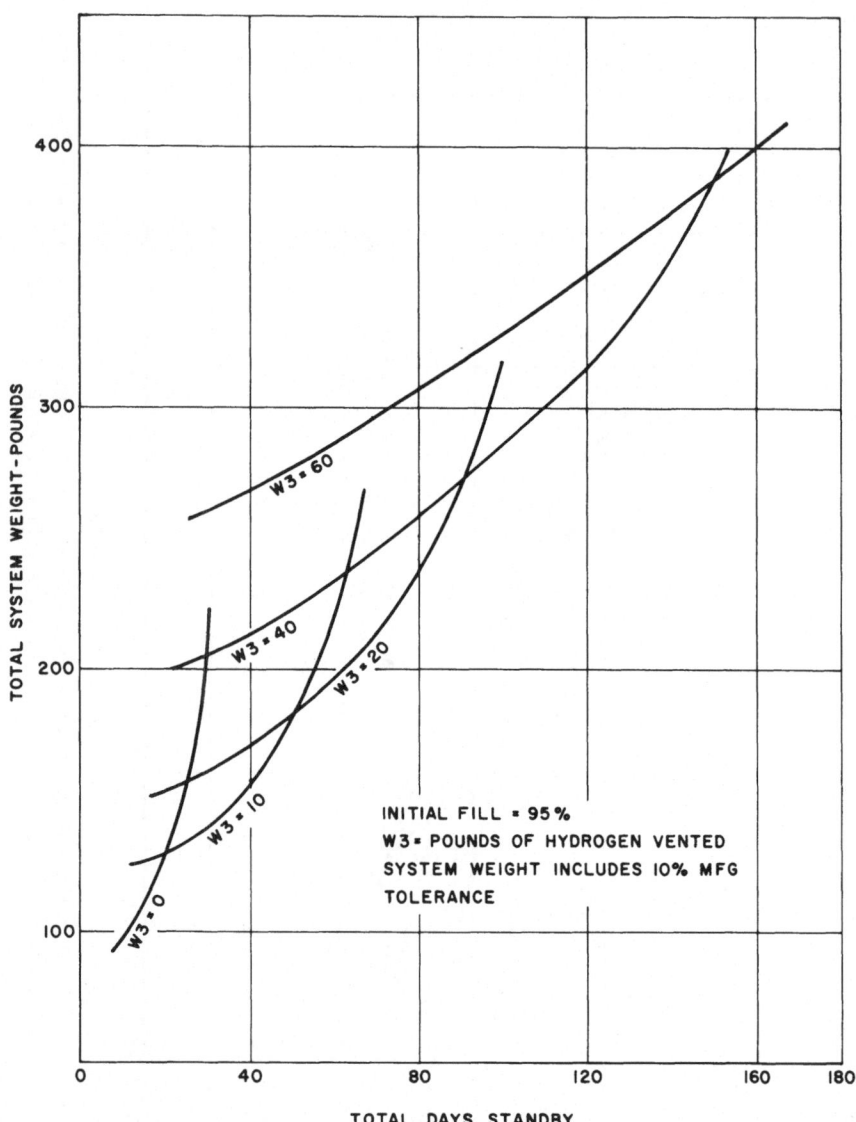

Figure 11. Subcritical cylindrical (L:D = 3) hydrogen system weight versus standby at 103 p.s.i.a. for 30 lb. of usable fluid. Initial fill, 95%, W3, pounds of hydrogen vented; 10% manufacturing tolerance included in system weight.

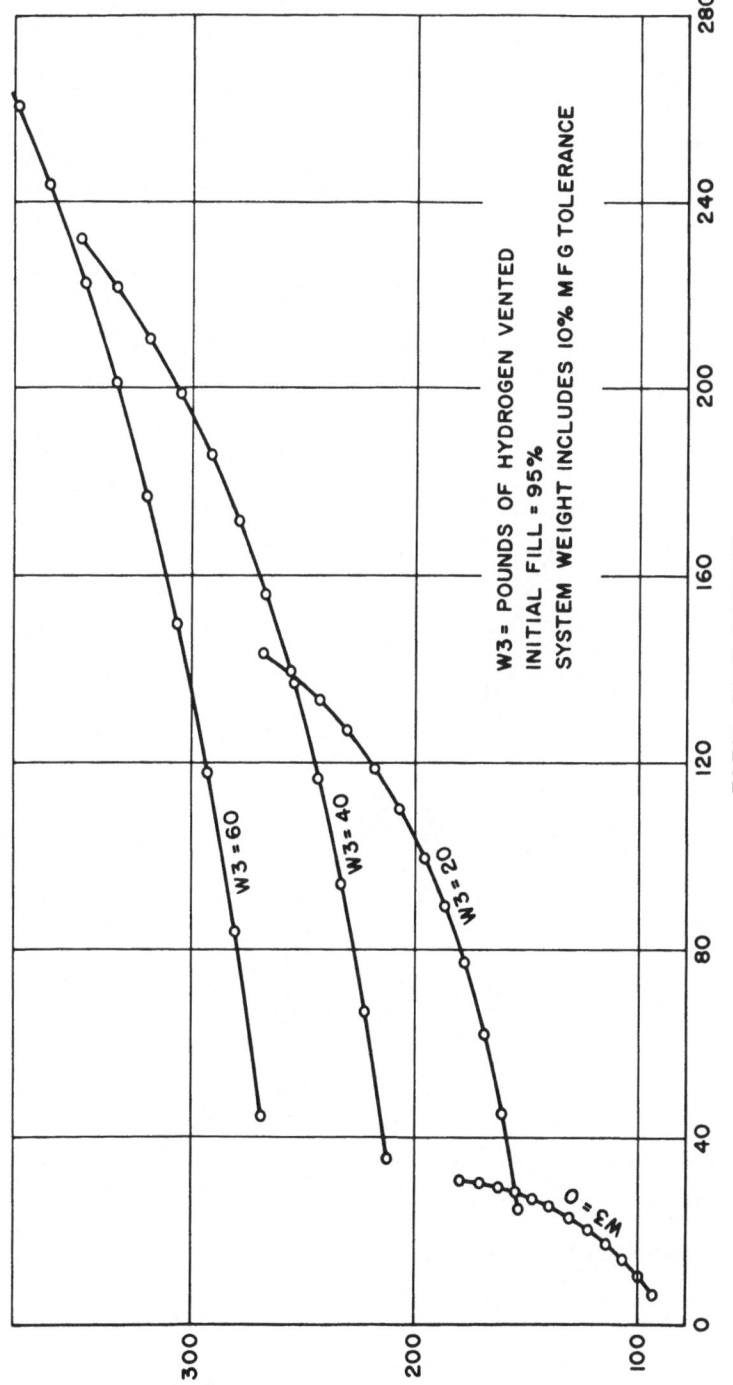

Figure 12. Supercritical spherical hydrogen system weight versus standby at 661.32 p.s.i.a. for 30 lb. of usable fluid. Initial fill, 95%; W3, pounds of hydrogen vented; 10% manufacturing tolerance included in system weight.

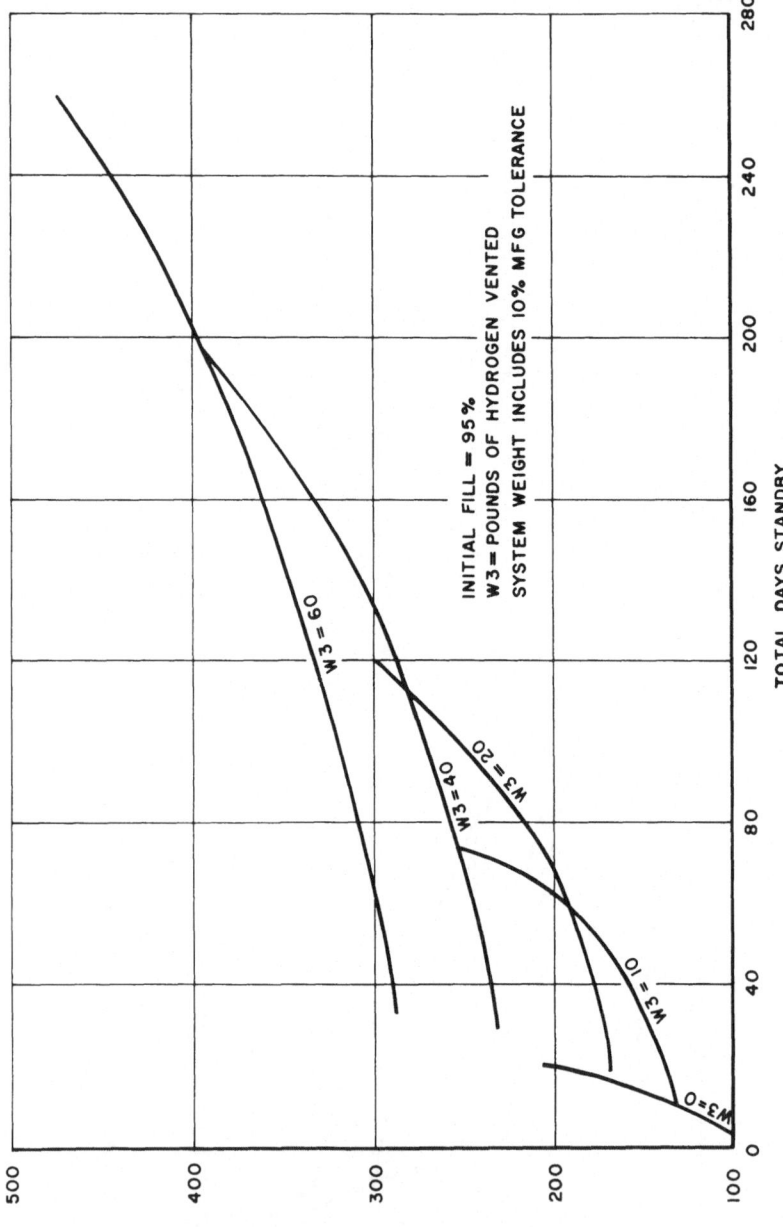

Figure 13. Supercritical cylindrical (L:D = 3) hydrogen system weight versus standby at 661.32 p.s.i.a. for 30 lb. of usable fluid. Initial fill, 95%; W3, pounds of hydrogen vented; 10% manufacturing tolerance included in system weight.

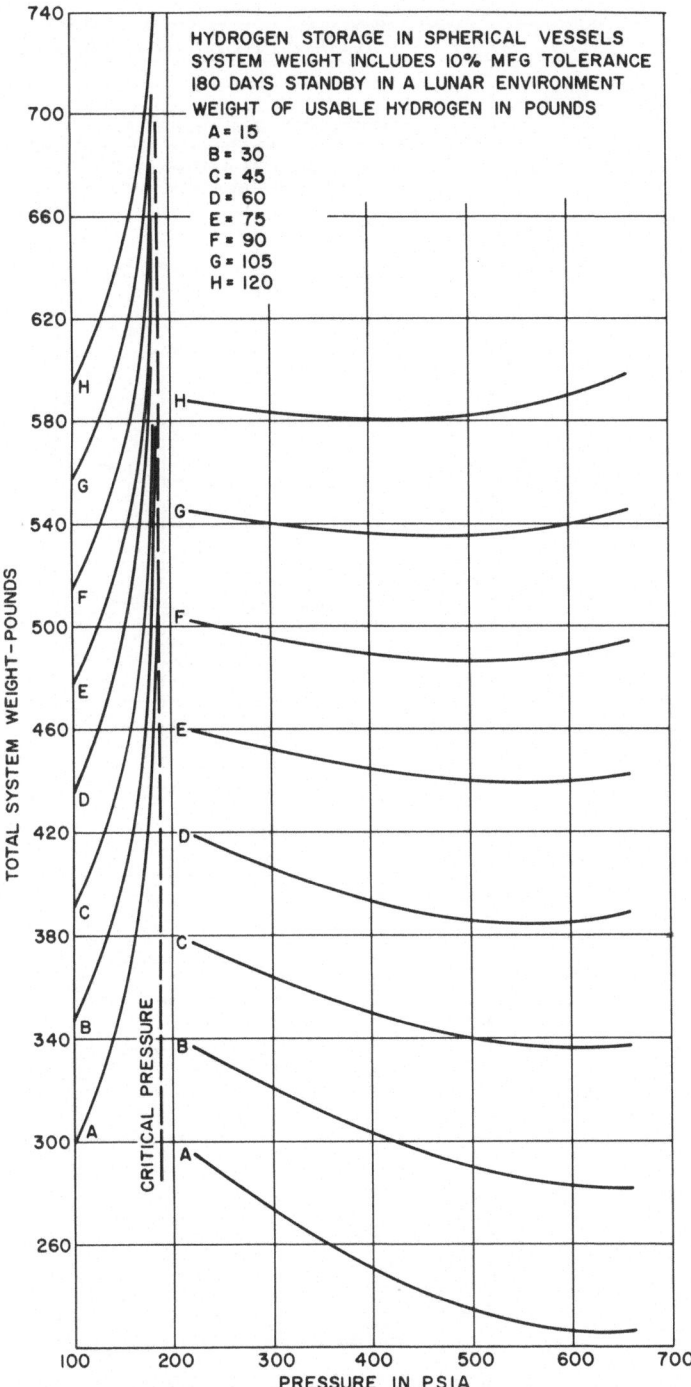

Figure 14. Minimum spherical hydrogen system weight versus pressure for 180 days standby. System weight includes 10% manufacturing tolerance. Weight of usable hydrogen in pounds: A, 15; B, 30; C, 56; D, 60; E, 75; F, 90; G, 105; H, 120;

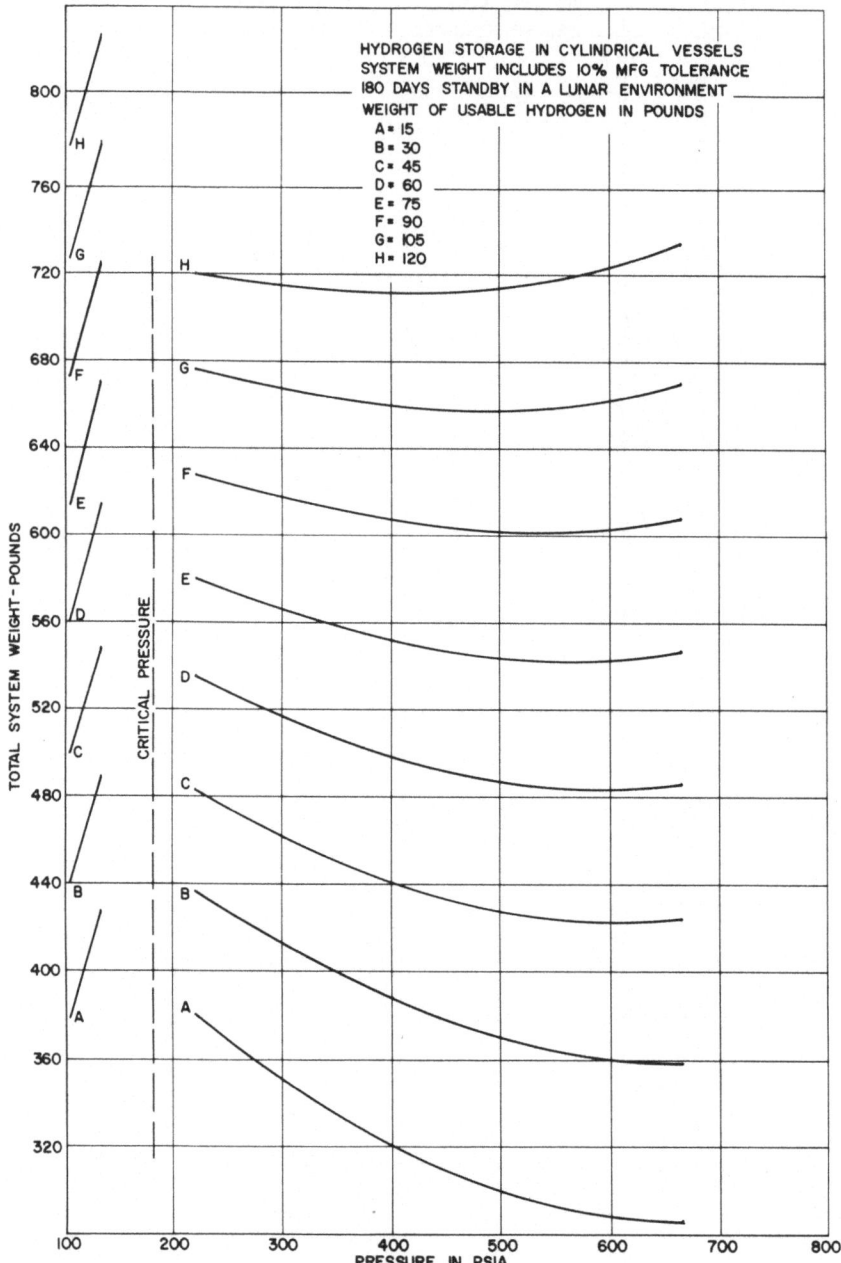

Figure 15. Minimum cylindrical (L:D = 3) hydrogen system weight versus pressure for 180 days standby. System weight includes 10% manufacturing tolerance. Weight of usable hydrogen in pounds: A, 15; B, 30; C, 45; D, 60; E, 75; F, 90; G, 105; H, 120.

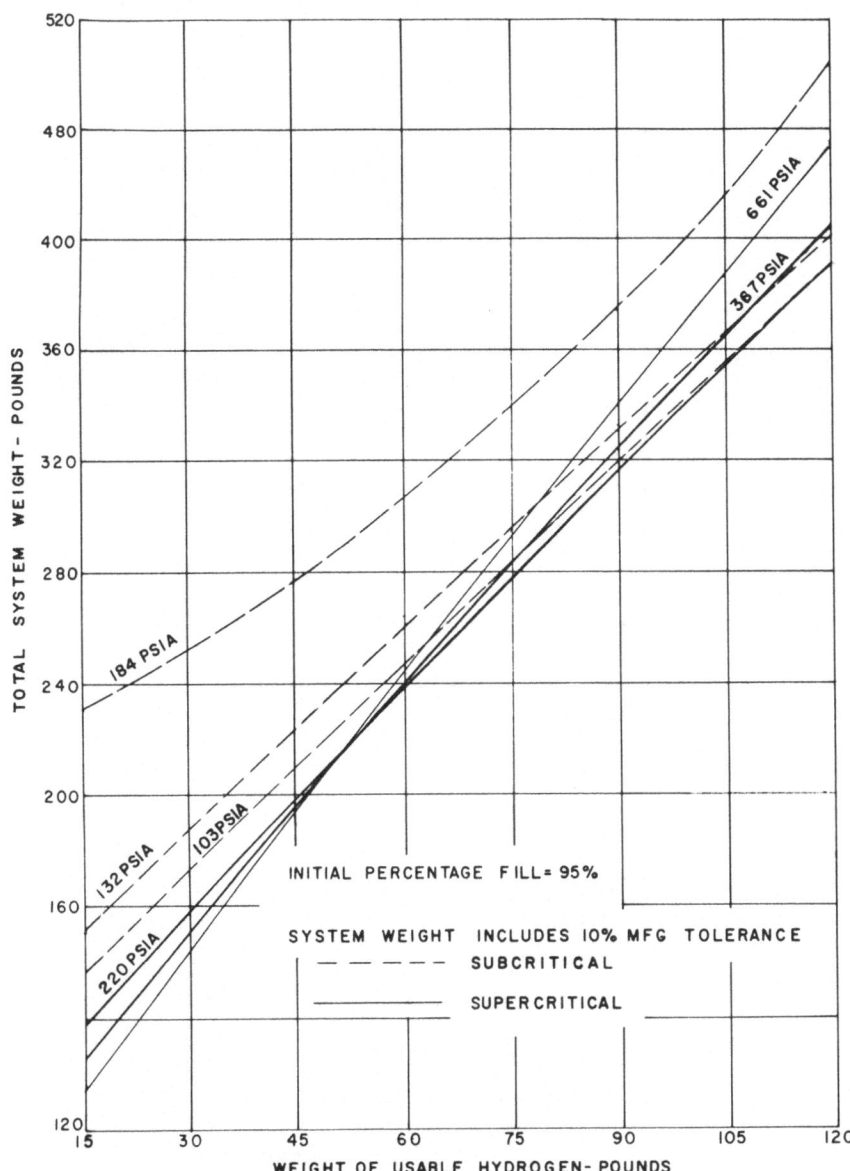

Figure 16. Minimum spherical hydrogen system weight (10% manufacturing tolerance included) versus usable fluid for 90 days standby. Initial
fill, 95%, broken line, subcritical pressure; solid line, supercritical pressure.

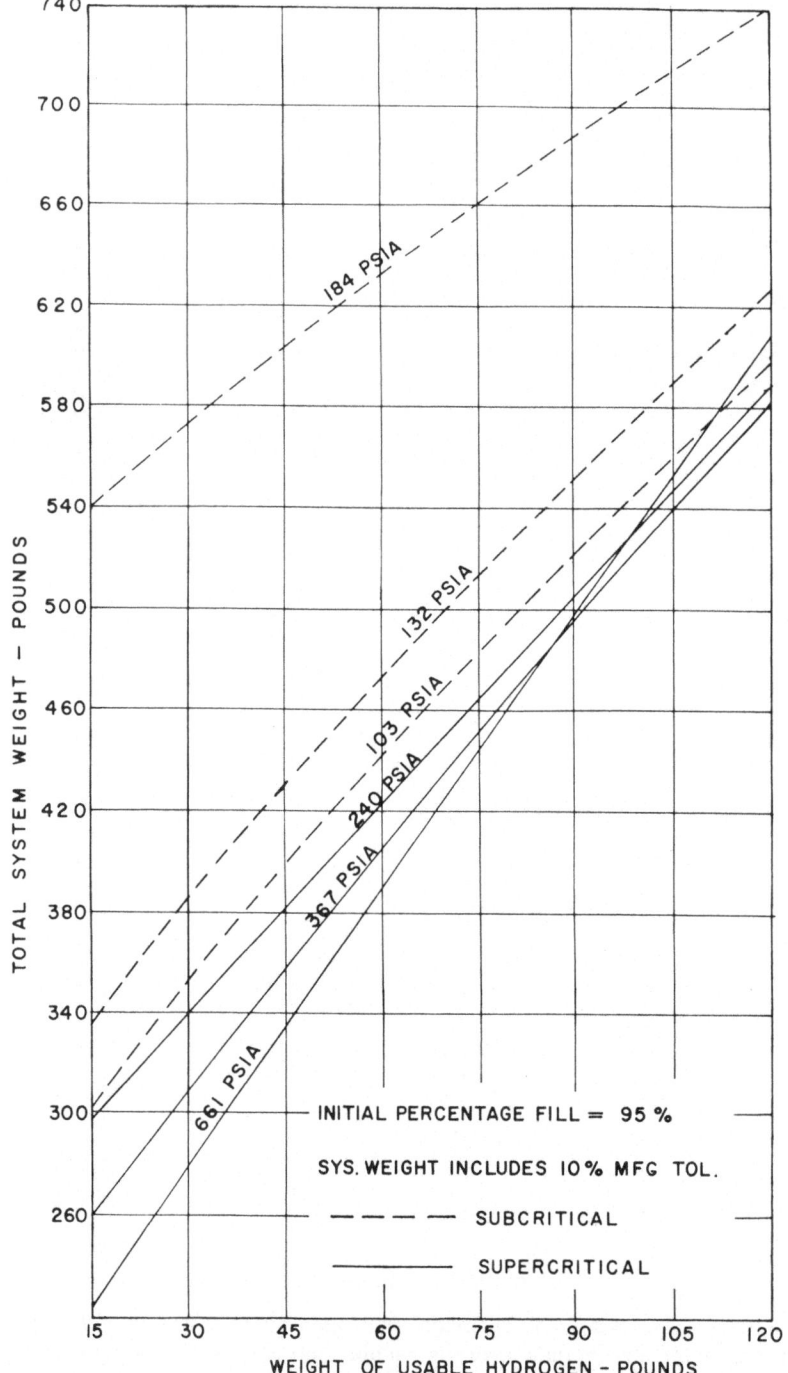

Figure 17. Minimum spherical hydrogen system weight (10% manufacturing tolerance included) versus usable fluid for 180 days standby. Initial fill, 95%, broken line, subcritical; solid line, supercritical.

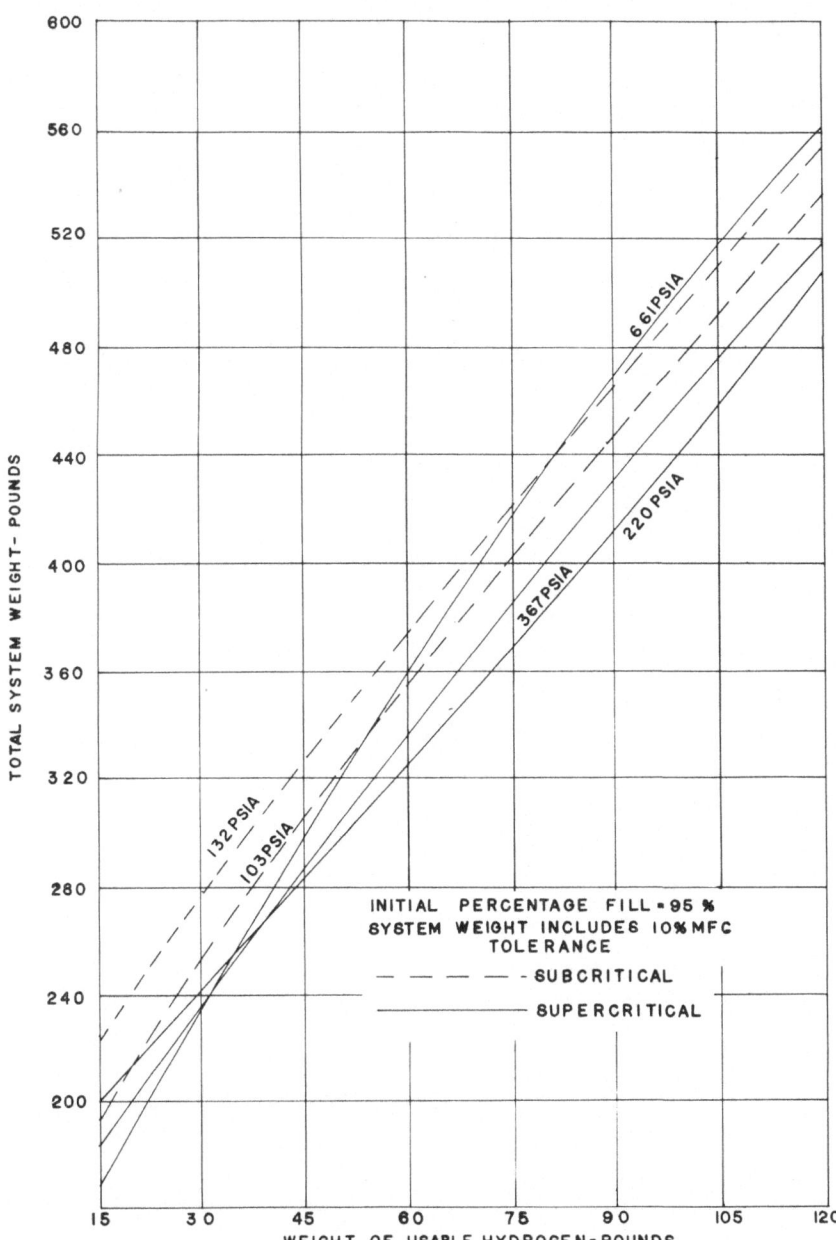

Figure 18. Minimum cylindrical (L:D = 3) hydrogen system weight (10% manufacturing tolerance included) versus usable fluid for 90 days stand-by. Initial fill, 95%; broken line, subcritical; solid line, supercritical.

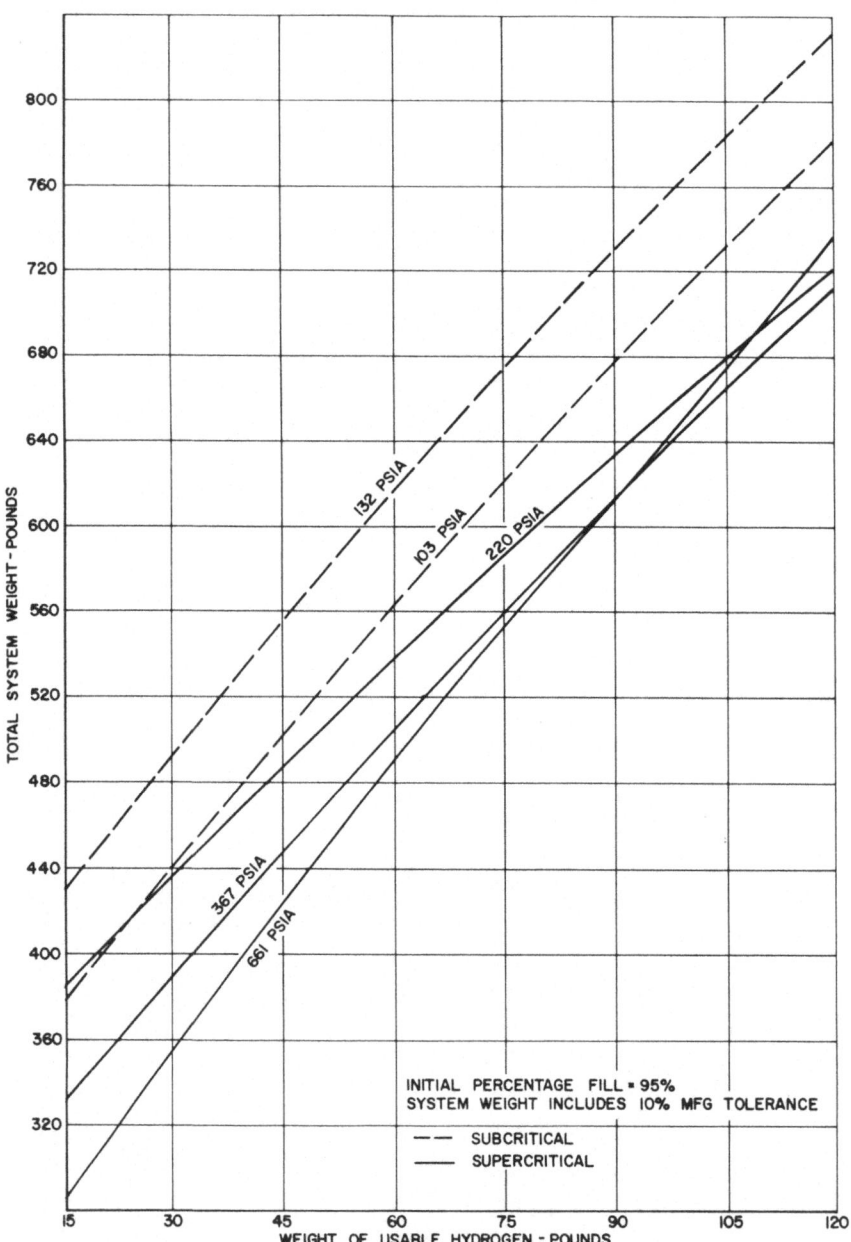

Figure 19. Minimum cylindrical (L:D = 3) hydrogen system weight (10% manufacturing tolerance included) versus usable fluid for 180 days standby. Initial fill, 95%; broken line, subcritical; solid line, supercritical.

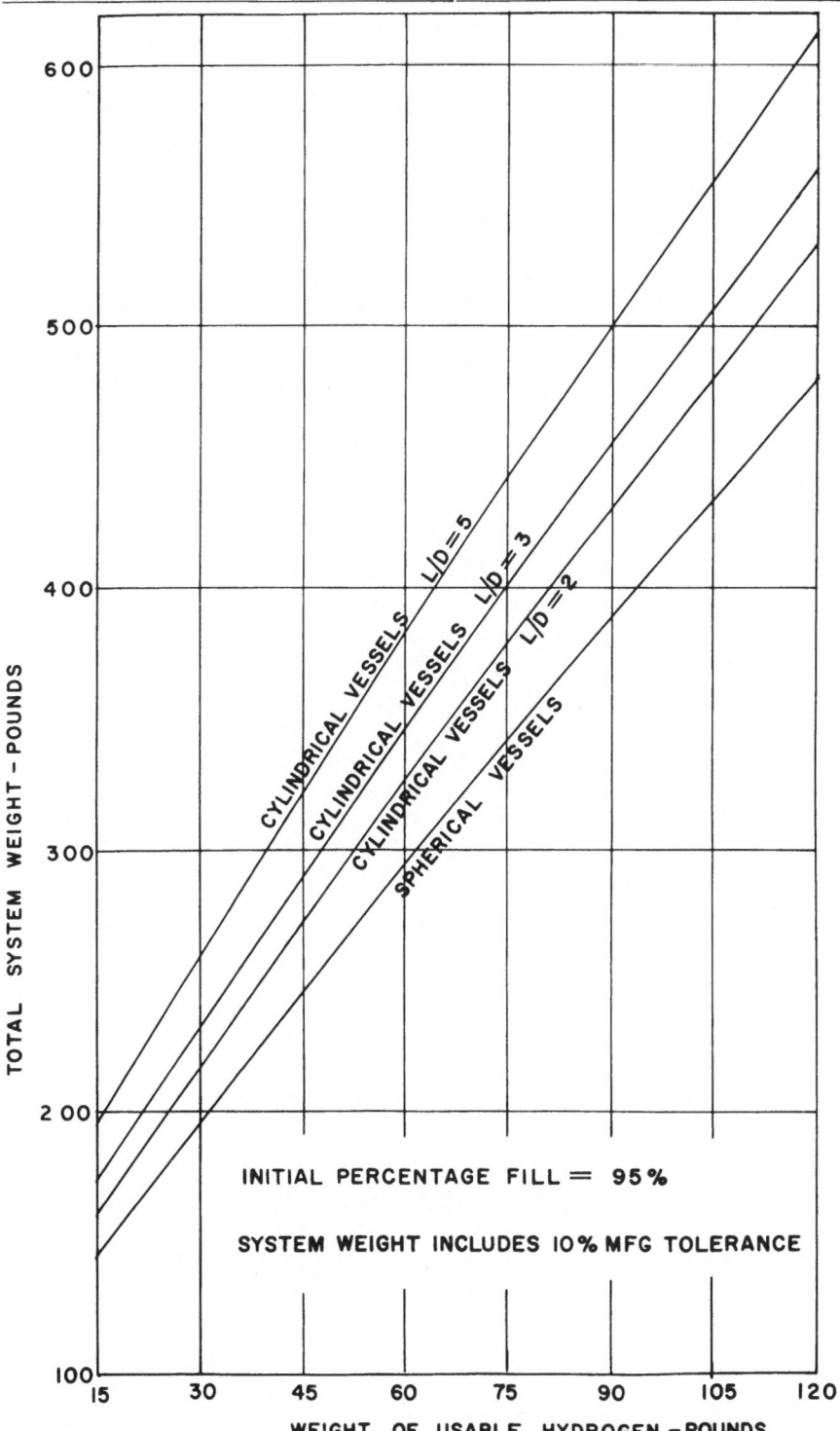

Figure 20. Minimum hydrogen system weight versus usable fluid and configurations at 661 p.s.i.a. for 90 days standby. Initial fill, 95%; 10% manu-

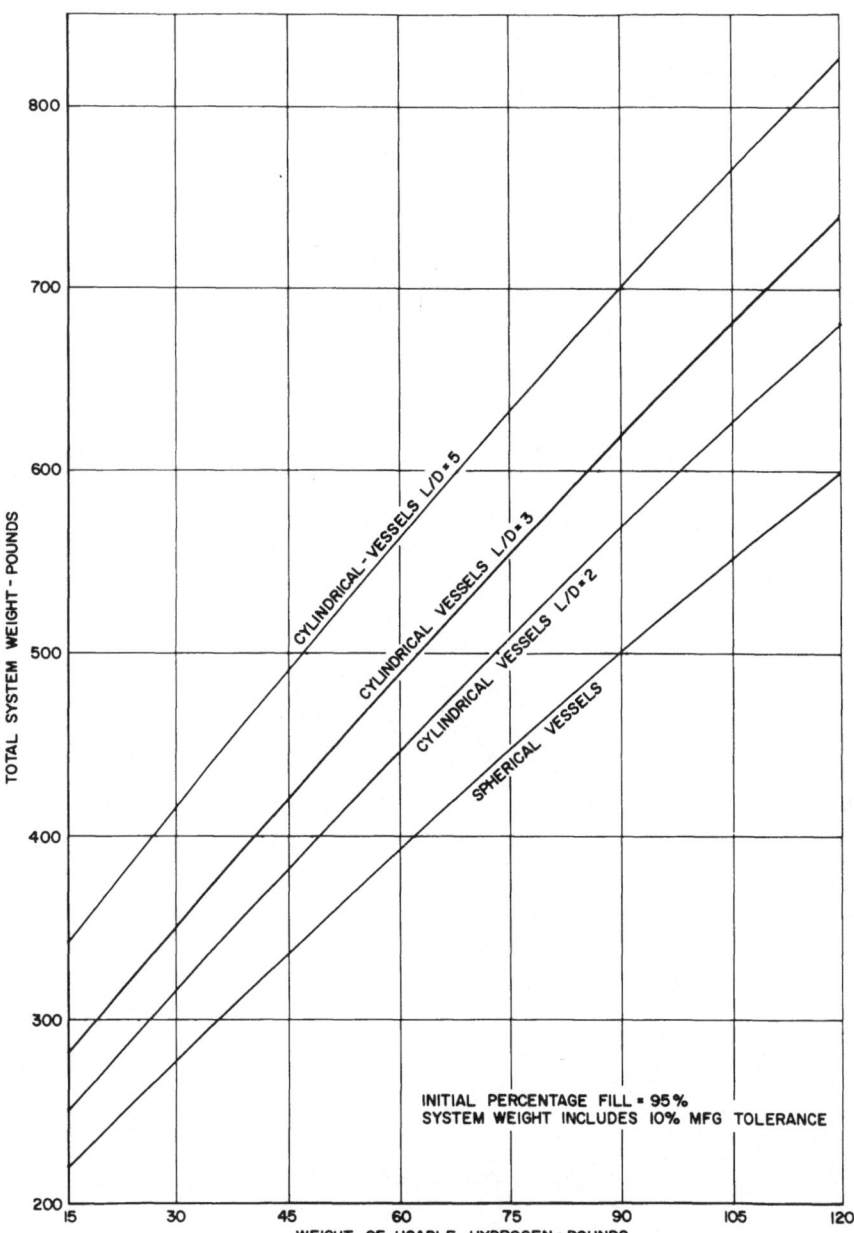

Figure 21. Minimum hydrogen system weight (10% manufacturing tolerance included) versus usable fluid and configurations at 661 p.s.i.a. for 180 days standby. Initial fill, 95%.

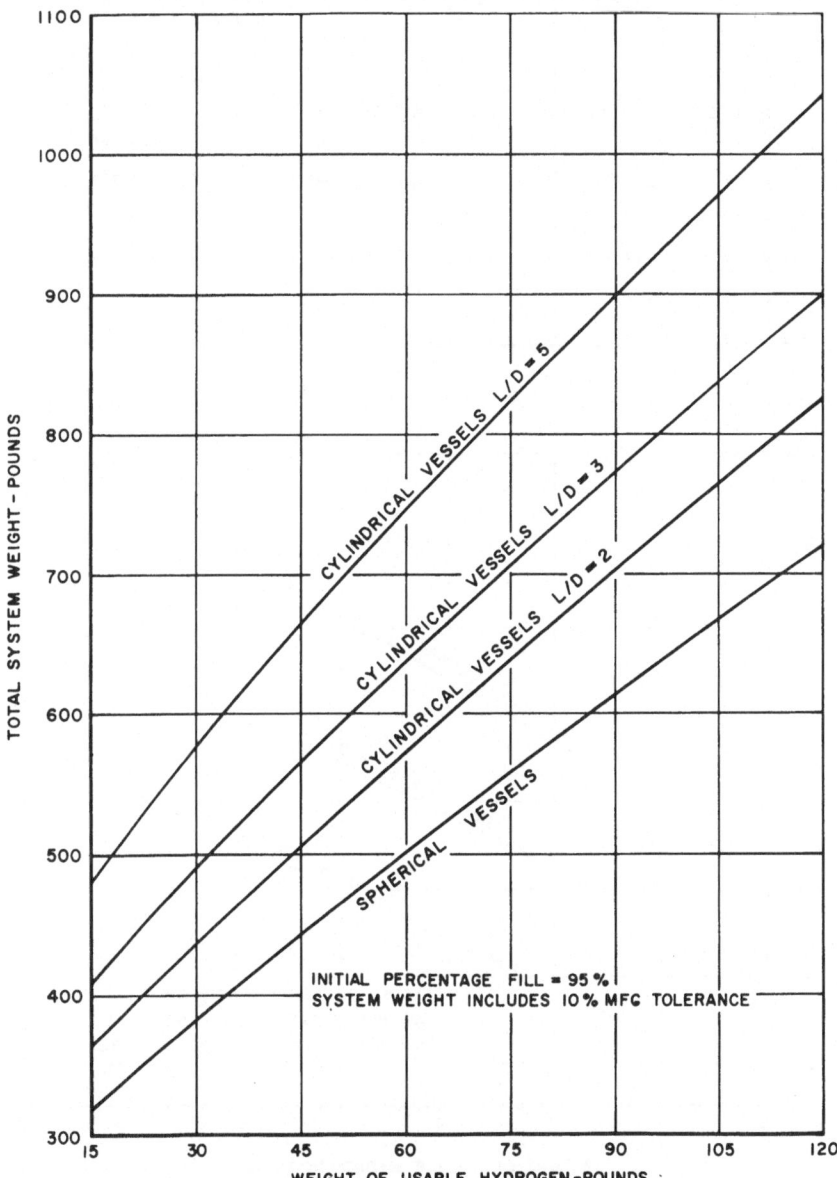

Figure 22. Minimum hydrogen system weight (includes 10% manufacturing tolerance) versus usable fluid and configurations at 661 p.s.i.a. for 270 days standby. Initial fill, 95%.

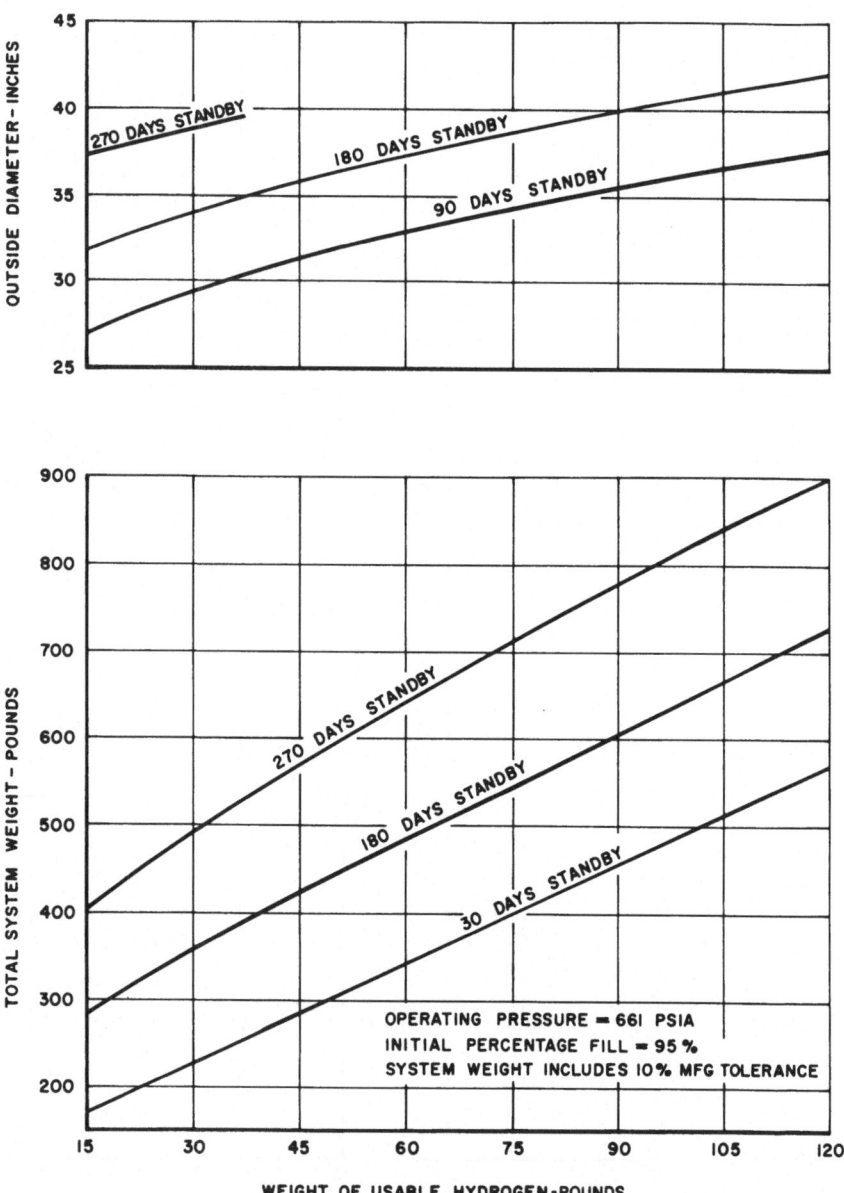

Figure 23. Minimum cylindrical (L:D = 3) hydrogen system weight (10% manufacturing tolerance included) and diameter versus usable fluid for 90, 180, and 270 days standby. Operating pressure, 661 p.s.i.a.; initial fill, 95%.

CONCLUSIONS

The general conclusions and trends from the available data follow. They are based on standby times varying from 90 to 270 days and on the range of usable fluid given.

Oxygen

1. Nonvented storage of liquid oxygen in spherical vessels yields minimum weight systems.
2. Subcritical pressure oxygen storage systems are lighter than supercritical pressure storage systems.
3. The optimum storage pressure for oxygen is dependent on the quantity of usable oxygen stored and the number of days of storage.
4. Decreasing the initial percentage fill for supercritical pressure storage of oxygen increases the days standby.

Hydrogen

1. Vented storage for hydrogen yields minimum weight systems.
2. Spherical storage systems are lighter than cylindrical storage systems.
3. Supercritical pressure hydrogen storage systems are lighter than subcritical storage systems.
4. The optimum supercritical storage pressure for hydrogen is dependent on the quantity of usable fluid being stored and the number of days of storage.
5. The higher initial percentage fill for hydrogen storage at supercritical pressure yields minimum weight systems.

AREAS REQUIRING REFINED ANALYSES

This preliminary investigation is based on certain simplifying assumptions; therefore, a number of refinements of this program are possible for the specific mission and advancements in technology. In addition, more specific and accurate data are required before a truly optimized system can result. The following are some of the factors that must be incorporated into a comprehensive optimization study, and some of the studies that must be made.

1. Trade-off studies using other materials, such as cryoformed stainless steel, for the oxygen and hydrogen pressure vessels.

2. Investigation of the effect of subcooling and slush.
3. Inclusion of vapor cooling in the vented standby analyses.
4. Investigation of the effect of discrete radiation shields, and inclusion of vapor cooling during vented standby.
5. Investigation of the integration of hydrogen vapor cooling on the oxygen storage systems.
6. Trade-off studies of monocoque outer shell structures, honeycomb or sandwich structure, and plastic bag or ground equipment cooling shrouds, including cooling of the internally mounted heat exchangers by ground supply fluid (liquid helium, etc.).
7. Investigation of the degree of stratification obtained by thermal conductors.
8. Refinement of the stratification assumptions by use of de-stratification devices in place of thermal conductors.
9. Inclusion of the effect of the *para* to *ortho* conversion in the hydrogen storage analyses.
10. Inclusion of the effect of the temperature on the thermal conductivity of the superinsulation.
11. Investigation of the effectiveness of superinsulation that is applied to compound curvature surfaces.
12. Refinement of the weight analyses for components that vary with vessel size and shape.
13. Investigation and inclusion of the concept of venting liquid instead of vapor in the subcritical storage systems.
14. Refinement of the heat input analysis by considering the increasing temperature of the cryogen during pressurization and during venting in supercritical storage systems.
15. Investigation and inclusion of the thermal lag through the superinsulation in the analyses.
16. Refinement of the heat leak through the bumper system (tank support system) to include the effect of varying the size of the bumpers.
17. Use of the equation for cylindrical buckling for the external cylindrical vacuum jacket.
18. Investigation and use of weight optimization procedures in the design of both pressure and vacuum jacket cylindrical sections.

DESIGN EXAMPLE

With the foregoing analysis as a basis, a number of refinements were incorporated in the determination of weight-optimized cryogenic storage and supply systems for a specific requirement. The cryogenic fluids required for the fuel cells of the power systems and the life-support function of the cabin systems are stored in double-walled evacuated Dewar-type vessels in which the evacuated storage vessels retain the required quantities of cryogen after 180 days of lunar standby. The cryogenic hydrogen system contains 63 lb. of usable hydrogen at the end of 180-day standby. The cryogenic oxygen system contains 721 lb. of oxygen.

Two tanks are provided for each cryogen to ensure a high reliability for emergency return and to facilitate vehicle design integration and center of gravity (cg) control. Each tank is a self-sufficient cryogenic storage system capable of independent operation. The cryogenic storage vessels are equipped with quantity, temperature, and pressure sensors in addition to the required components for monitoring the status and performing the operational requirements imposed on the system.

The optimum system weight to store oxygen for 180 days is obtained in a nonvented system operating below the critical pressure in the subcritical state. The oxygen is stored in two identical nonvented spherical vessels at a subcritical pressure of 294 p.s.i.a. This two-phase system will operate between the storage pressure and 100 p.s.i.a. and will deliver either liquid oxygen, or a gas mixture, or both, to the fuel cells and life-support system.

Oxygen system

The design basis requirements for the oxygen system follow; the design features are presented in Table 5.

1. Subcritical storage nonvented standby.
2. Two spherical storage vessels.
3. 360.5 lb. of usable oxygen per tank.
4. 180-day standby.
5. Maximum storage pressure 294 p.s.i.a.
6. A residual density of 0.9 lb./cu. ft.
7. Linde SI-44 superinsulation having an effectiveness of 50%.
8. Cryoformed stainless steel for the inner vessel with a minimum thickness of 0.015 in.
9. Aluminum 6061T-6 for the outer vessel.
10. 100% fill at the maximum storage pressure of 294 p.s.i.a.

Hydrogen is stored in two vented cylindrical vessels designed and initially filled to obtain a supercritical pressure of 661 p.s.i.a. prior to venting and prior to completion of 180-day standby. The operating pressure of the supercritical system during fluid withdrawal will range from 661 to 100 p.s.i.a.

TABLE 5

OXYGEN SYSTEM DESIGN FEATURES

Storage Vessel Data	
External diameter (in.)	30.76
Internal diameter (in.)	27.97
Inner vessel wall thickness (in.)	0.0150
Outer vessel wall thickness (in.)	0.0406
Insulation thickness (in.)	1.339
Initial percentage fill (%)	77.7
Total fluid weight per tank (lb.)	366.47
Dry system weight per tank (lb.)	91.20
Total system weight per tank (lb.)	457.67
System Weights (lb.)	
Dry weight per tank:	
Inner vessel	11.62
Insulation	10.85
Outer vessel	12.98
Valves	8.63
Supports and Miscellaneous	47.12
Total	91.20
Fluid weight per tank:	
Usable at end of 180-day standby	360.50
Vented fluid	0.00
Residual fluid (unusable)	5.97
Total	366.47
Total weight per tank	457.67
Total system weight (two tanks)	915.34

Hydrogen system

The summarized design basis requirements for the hydrogen system follow; the design features are summarized in Table 6.

1. Supercritical storage with vented standby.
2. Two cylindrical storage vessels.
3. 31.5 lb. of usable hydrogen per tank.
4. 180-day standby.
5. Maximum storage pressure of 661 p.s.i.a.
6. A residual density of 0.35 lb./cu. ft.
7. Linde SI-44 superinsulation having an effectiveness of 50%.
8. Cryoformed stainless steel for the inner vessel with a minimum thickness of 0.015 in.

9. Aluminum 6061T-6 for the outer vessel.
10. Aspect ratio of 3 for the cylinder based on the internal diameter of the inner vessel.

TABLE 6

HYDROGEN SYSTEM DESIGN FEATURES

Storage Vessel Data	
External diameter (in.)	30.44
Internal diameter (in.)	22.97
Inner vessel wall thickness (in.)	
Spherical section	0.0222
Cylindrical section	0.0444
Outer vessel wall thickness (in.)	
Spherical section	0.0400
Cylindrical section	0.1070
Insulation thickness (in.)	3.605
Initial percentage fill (%)	95.0
Total fluid weight per tank (lb.)	61.64
Dry system weight, per tank (lb.)	244.67
Total system weight per tank (lb.)	306.31
System Weights (lb.)	
Dry weight per tank:	
Inner Vessel	57.92
Insulation	65.37
Outer Vessel	62.90
Valves	8.63
Supports and Miscellaneous	49.85
Total	244.67
Fluid weight per tank:	
Usable at end of 180-day standby	31.50
Vented fluid	25.00
Residual fluid (unusable)	5.14
Total	61.64
Total weight per tank	306.31
Total system weight (two tanks)	612.62

FURTHER IMPROVEMENTS

The design of the cryogenic storage system was optimized using the techniques indicated. However, improvements in heat leak and performance will be possible by refining or altering these techniques and assumptions.

In general, a reduction in heat leak is possible by increasing the amount of superinsulation at the expense of an increase in weight. Also, as already indicated, the use of a venting-type standby can reduce the insulation requirement for certain fluids and mission situations. By incorporating a vapor cooling technique, the venting standby can be enhanced.

Vapor cooling

An extension to the venting standby procedure is the function of vapor cooling of the insulation. Wherever a flow situation exists (that is, a flow of the cold cryogenic fluids from the tank) the mechanics for vapor cooling of the insulation exists. The procedure is to feed the fluid supply line through the insulation layers in the form of an expansion heat exchanger to cool the insulation. A portion of the environmental heat absorbed by the insulation is absorbed and carried from the tank by the fluid supply, thus minimizing the heat input to the storage tank.

The use of vapor cooling can either extend the standby time or reduce the insulation weight required for any critical heat input venting or flow situation. Vented, vapor cooling is applicable to either supercritical or subcritical storage systems. Vapor cooling applied to the present nonventing cryogenic storage systems may show a weight reduction. The use of vapor cooling during the filling operation, moreover, will precool the superinsulation, which will further reduce the radiation and conduction heat transfer during the period that the vessel is in the pressurization process.

The degree of improvement that can be expected using vapor cooling in conjunction with superinsulation at this time is not known. From the meager data available on this subject, it is estimated that a 50% reduction in the thickness of insulation is possible during the venting phase. However, the effectiveness of the insulation assumed in the preliminary design is even at this time questionable, as present cryogenic tankages now being built for space programs have not achieved the effective thermal conductivity used in these analyses. Thus, considerable data must be collected to establish attainable effective thermal conductivities and the effect of vapor cooling on these values.

Discrete radiation shields

For nonvented cryogenic storage systems, the present technology in supporting discrete radiation shields does not lend itself favorably for long periods of standby. However, present testing has shown that the use of discrete radiation shields in even these conditions will minimize the radiation and conductive heat leaks in a cryogenic system. Furthermore, these tests have indicated that a more predictable heat leak can be attained using discrete radiation

TABLE 7

DISCRETE RADIATION SHIELD SUMMARY[a]

Volume (cu. ft.)	Total fluid (lb.)	Vented weight (lb.)	No. of days venting
21.18	88.91	50	313
15.98	67.10	30	225
13.38	56.18	20	169
10.79	45.28	10	98

[a] From these data, the 180-day standby requirement could be met by a cylindrical vessel venting 20 lb. of fluid. The total system weight was determined to be 274.2 lb., or a reduction of 32.1 lb. for each hydrogen storage system. In addition, a further reduction of about 5.5 in. in the diameter of the vessel is realized.

shields than with superinsulation. A greater advantage is realized with discrete radiation shields that are isothermally mounted to the annular tubes when a vented standby technique is used and vapor cooling can be employed.

An analysis was made for a cryogenic hydrogen storage system meeting the requirements previously outlined. A cylindrical vessel having an aspect ratio (L:D) of 3 and four discrete radiation shields, with the innermost shield vapor cooled, was selected. For this analysis, the outer surface temperature was assumed to be 13°F. for the first 7 days, then alternated from −136.6 to 13°F. for subsequent 14-day periods. Incorporated in the analyses were the heat input through the inner vessel support system as well as factors for the thermal conductivities of the shield support mechanisms.

Table 7 summarizes this analysis and shows the number of days of vented standby achievable for a hydrogen system storing 31.5 lb of usable fluid and having four discrete radiation shields, one of which is a vapor-cooled shield.

REFERENCES

Roder, H. M., L. A. Weber, and R. D. Goodwin (1963). The thermodynamic functions of parahydrogen from the triple point to 100°K at pressures up to 340 atmospheres. NBS Rept. 7639, Cryogenic Engineering Lab., Boulder, Colorado (January).

Stewart, R. B., J. G. Hust, and R. D. McCarty (1963). Interim thermodynamic properties for gaseous and liquid oxygen at temperatures from 55 to 300°K and pressures to 300 atmospheres. NBS Rept. 7922, Boulder, Colorado (October).

4.

AN ELECTROLYTIC PROCESS FOR CARBON DIOXIDE SEPARATION AND OXYGEN RECLAMATION

W. E. ARNOLDI

Hamilton Standard, Division of United Aircraft Corporation
Windsor Locks, Connecticut

OXYGEN RECLAMATION FROM CARBON DIOXIDE

General requirements

For the manned space vehicle whose mission duration without resupply is several months or more, it is generally anticipated that it will be logistically worthwhile, if not absolutely essential, to derive a large part of the oxygen required for respiration from the carbon dioxide exhaled. It is the purpose of this paper to describe and discuss, at a fundamental level, one particular method for performing this function, providing a basis for rational quantitative treatment of the chemical and electrochemical processes involved, as inferred from experimental research currently in progress. Interpretation of the present state of knowledge in terms of elementary principles of chemical equilibrium is emphasized; where support of assumptions or simplifications is necessary, experimental observations are mentioned.

The ideal oxygen reclamation system would perform four basic functions:

1. Separation of carbon dioxide from the space cabin atmosphere, whose total pressure would be predetermined at some level between 160 mm. Hg and 760 mm Hg, carbon dioxide partial pressure being between 4 and 8 mm. Hg;

2. Decomposition of carbon dioxide to its elementary constituents, solid carbon and gaseous oxygen;

3. Delivery of oxygen to the cabin atmosphere;

4. Delivery, for disposal or storage, of carbon.

In practice, it may not be essential to produce and deliver pure carbon or even elementary carbon, but any additional material to be disposed of as a contaminant or as part of a carbon compound must be limited, since such expendable mass directly detracts from the advantages of oxygen reclamation.

The process described in this paper is potentially capable of closely approaching the ideal requirements in a single unit of equipment, removing carbon dioxide and replenishing it with oxygen simultaneously, while collecting solid carbon for periodic disposal. Since energy consumption in a space vehicle also represents a logistic penalty, it is significant that the electrical efficiency of the process described is high.

Basic process concept

Decomposition of carbon dioxide into solid carbon and molecular oxygen may be accomplished by means of concurrent chemical and electrochemical processes in certain fused carbonate electrolytes. A program of experimental and theoretical research undertaken at Hamilton Standard, currently sponsored by the NASA Langley Research Center under contract NAS 1-4154, has shed considerable light on the nature of these processes, providing support for working hypotheses from which system performance may be evaluated and engineering specifications laid down.

The basic configuration for this system is indicated by the diagram of Figure 1, showing a cell in which carbon dioxide is absorbed from a gas stream by direct contact with the electrolyte, oxygen is evolved at the anode, and carbon is accumulated in solid form at the cathode. The required operating temperature is maintained by energy losses in the process and by thermal insulation. Thermal losses to the effluent air are minimized by the use of a regenerative heat exchanger between inflow and outflow.

The fundamental concept is that lithium, electrochemically reduced at the cathode in an electrolyte containing lithium carbonate, reacts chemically with carbon dioxide in solution to deposit solid carbon adherent to the cathode surface, returning lithium ions and oxide ions to the electrolyte. At the anode, oxide ions are discharged to form gaseous oxygen. Since three oxide ions are produced at the cathode for each pair of oxide ions discharged at the anode, the extra ion is available to absorb a molecule of carbon dioxide gas, replenishing, as carbonate, the carbon dioxide reduced at the cathode. There is no substantial accretion of lithium at the cathode, since lithium atoms are promptly returned to ionic form by reaction with carbon dioxide. However, a number of secondary

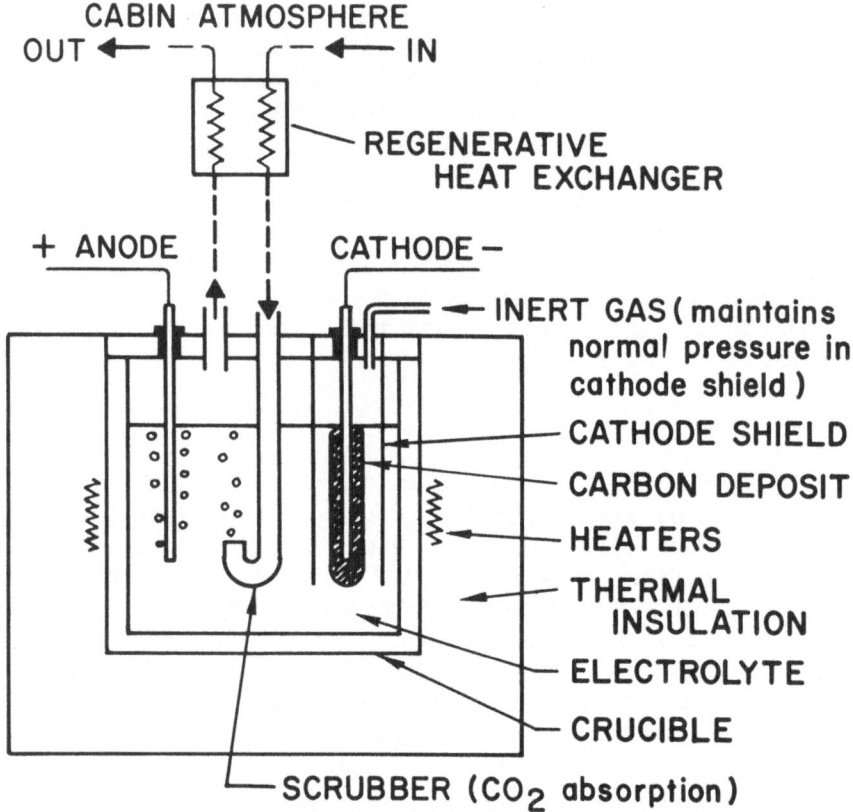

Figure 1. Schematic of carbon dioxide electrolytic cell.

reactions may take place, and are indeed required for chemical equilibrium; hence, a more complete exposition of the concept will be presented in order to indicate constraints in the application to an engineering process.

Notation or terminology that is peculiar to this paper is explained where first used in the text. Table 1, however, has been prepared to permit convenient reference to definitions of equilibrium constants and gas partial pressure symbols that are repeatedly used.

Reactions at the electrodes

Consider first an electrolyte comprising a eutectic mixture of lithium carbonate and lithium chloride. This composition is not essential, but it has a convenient melting point (509°C.), is chemically stable under a small partial pressure of CO_2, and will provide a reasonably simple basis for discussing fundamental requirements in CO_2 decomposition. In the molten state—say, at 550°C.—this

TABLE 1

SPECIAL SYMBOLS AND NOMENCLATURE

1. EQUILIBRIUM CONSTANTS

Symbol	Reaction	Eq. no.
K_0	$4Li + CO_2 \rightarrow 4Li^+ + 2O^{2-} + C$	(2)
K_1	$CO_2 + O^{2-} \rightleftharpoons CO_3^{2-}$	(14)
K_2	$H_2O + O^{2-} \rightleftharpoons 2OH^-$	(21)
K_3	$C + CO_2 \rightleftharpoons 2CO$	(20)
K_4	$C + 2H_2O \rightleftharpoons 2H_2 + CO_2$	(29)
K_5	$C + 2H_2 \rightleftharpoons CH_4$	(30)

2. GAS PARTIAL PRESSURES

Symbol	Gas
p	Carbon dioxide
v	Water vapor
c	Carbon monoxide
h	Hydrogen
m	Methane

mixture will be highly ionized. Lithium will be the only cation, but anions will include carbonate, chloride, and oxide that results from the dissociation of carbonate ions. Under equilibrium conditions, the dissociation of carbonate anions will be determined by the ionic activities and by the partial pressure of CO_2 at a free surface. Although lithium oxide and lithium chloride will both have vapor pressures, these are small and may be ignored for the purpose of this discussion.

At the cathode, lithium ions will be reduced in accordance with Eq. 1.

$$4Li^+ + 4e \rightarrow 4Li \qquad (1)$$

The atomic lithium will react with CO_2, available by dissociation of carbonate ions, to yield elementary carbon.

$$4Li + CO_2 \rightarrow 4Li^+ + 2O^{2-} + C \qquad (2)$$

It is frequently convenient to represent the net result of these cathode reactions by summation:

$$CO_2 + 4e \rightarrow C + 2O^{2-} \tag{3}$$

However, it should be noted that this is merely symbolic for the net cathodic process; the distinction between the *electrochemical* reaction of Eq. 1 and the *chemical* reaction of Eq. 2 should not be ignored.

At the anode, oxide ions will be oxidized in accordance with Eq. 4.

$$2O^{2-} \rightarrow O_2 + 4e \tag{4}$$

Thus, the net *electrochemical* process, represented by Eqs. 1 and 4, is the decomposition of lithium oxide:

$$4Li^+ + 2O^{2-} \rightarrow 4Li + O_2 \tag{5}$$

The potential difference at equilibrium between an anode and a cathode which are themselves chemically inert may be expressed by Eq. 6, which is the summation of the Nernst equations for the two electrodes.

$$E = E_0 - \frac{RT}{4F} \ln \frac{(Li^+)^4 (O^{2-})^2}{(Li)^4 (O_2)} \tag{6}$$

E_0 is the potential difference for unit activity of reactants and products, determined by the free energy change, ΔF_1, in Eq. 5.

$$E_0 = \Delta F_1/4F = 2.5 \text{ volts at } 550°C. \tag{7}$$

The concentration terms in Eq. 6 may be replaced by more conveniently available quantities by considering equilibrium of the chemical reaction at the cathode, Eq. 2. The equilibrium constant may be expressed by

$$K_0 = \frac{(Li^+)^4 (O^{2-})^2 (C)}{(Li)^4 (CO_2)} \tag{8}$$

whence, by substitution into Eq. 6, we have

$$E = E_0 - \frac{RT}{4F} \ln \frac{(CO_2)K_0}{(O_2)(C)} \tag{9}$$

Since K_0 is definable in terms of the free energy change, ΔF_0, for the reaction of Eq. 2, substitution into Eq. 9 yields

$$E = \frac{\Delta F_1 + \Delta F_0}{4F} - \frac{RT}{4F} \ln \frac{(CO_2)}{(O_2)} \tag{10}$$

where (C) has been set to unity, since only solid carbon will be present. Since adding Eqs. 2 and 5 results in decomposition of CO_2 as a net process, the sum of the two free energy changes, $\Delta F_1 + \Delta F_0$, corresponding to these reactions, is identical with the free energy change for the decomposition of CO_2 to carbon and oxygen. This free energy change thus determines the electrode potential difference for unit activity of oxygen and carbon dioxide, and the cell voltage may be written as

$$E = 1.025 - \frac{RT}{4F} \ln \frac{(CO_2)}{(O_2)} \tag{11}$$

Since the free energy of formation of carbon dioxide is substantially constant over a wide temperature range, this equation is not restricted to the 550°C. temperature proposed for this discussion. The activity of oxygen, (O_2), and of carbon dioxide, (CO_2), may be represented by the partial pressures of these gases.

EFFICIENCY OF THE ELECTROLYTIC CELL

In the application of a process for oxygen reclamation to space vehicle requirements, it is appropriate to investigate the efficiency achievable. Efficiency will be defined, for this purpose, as the quotient of ideal electrolytic power input divided by actual electrolytic power input, and the ideal power input will be based upon the free energy of formation of carbon dioxide. For an electrochemical process, in which Faraday's principle applies, the current will be determined uniquely by the gram-equivalent rate of decomposition, and the power input will be the product of applied potential by current. Laboratory experience has indicated that, for electrolytic decomposition of CO_2 in an alkali carbonate melt, 100% current efficiency is readily achieved; hence, the power is proportional to the voltage required. Thus, 100% power efficiency may be associated with an electrode potential difference of 1.025 volts, in accordance with the definition proposed, and increases in potential due to electrode over-voltages, concentration gradients, Joule heating losses, etc., will specify operating efficiencies, generally less than 100%.

For example, consider the idealized case where the cell delivers oxygen at a partial pressure of 0.20 atm., absorbing CO_2 at an input partial pressure of 0.005 atm., the process taking place at 550°C. Eq. 11, with these numerical values, becomes

$$E = 1.025 - \frac{8.31 \times 823}{4 \times 96500} \ln \frac{0.005}{0.020} = 1.090 \text{ volts} \tag{12}$$

For these circumstances, the limiting efficiency will be 1.025/ 1.090 or 94%. In practice, this level is unattainable, since the electrolyte must have a CO_2 partial pressure that is substantially less than the input CO_2 pressure in order that effective chemical absorption take place. If we arbitrarily set a 90% absorption efficiency as a requirement and assume that equilibrium is attained before the gas leaves the melt, the CO_2 pressure of the melt will be 0.0005 atm., resulting in a cell potential of 1.132 volts and an efficiency of 90.6%. Again, this becomes a target that cannot be closely approached in practice, since it is based on an equilibrium that is significantly disturbed when electrode current densities are not vanishingly small.

At the anode, depletion of oxide ions by discharge of gaseous oxygen gives rise to a local gradient in oxide ion concentration wherein electrolytic migration is balanced against diffusion processes that tend to restore equilibrium. Likewise, at the cathode, where oxide ions are generated in accordance with Eq. 2, a balance will be attained between the rate of oxide generation and the rate of diffusion away from the cathode, resulting in a locally decreased concentration of CO_2. The consequent difference in oxide ion concentration, or CO_2 concentration, between anode and cathode will further increase the cell potential required to sustain the electrolytic process in steady state.

We shall evaluate this·difference by considering two carbon cathodes, one of which is subject to the environment previously assumed while the other has an increased oxide ion concentration in its immediate vicinity. The difference in potential between these two carbon electrodes may then be added to the potential difference of Eq. 11 in order to include the effect of the change in oxide ion concentration from the anode region to the cathode. This approach is based on the assumption that Eqs. 1 and 2 represent reversible reactions at a carbon electrode, an assumption that is in accord with laboratory experience.

Using subscripts 1 and 2 to designate the two carbon electrodes and their associated environments, and representing CO_2 partial pressures by p_1 and p_2, the equation for the potential difference, $E_2 - E_1$, is written as

$$E_2 - E_1 = -\frac{RT}{4F} \ln \frac{p_2 (O^{2-})_1^2}{(O^{2-})_2^2 p_1} \tag{13}$$

In order to eliminate the oxide ion activities, or molal concentrations, from this equation, consider the equilibrium of oxide, CO_2 and carbonate:

$$CO_2 + O^{2-} \rightleftharpoons CO_3^{2-}, \quad K_1 = \frac{(CO_3^{2-})}{(O^{2-})p} \tag{14}$$

Since the total number of lithium ions and of chloride ions is fixed, the total number of electrochemical equivalents of oxide ions plus carbonate ions is also fixed, regardless of CO_2 pressure, each oxide ion being directly convertible to one carbonate ion, and vice versa. Their total molal concentration is thus constant and may be expressed as

$$(CO_3^{2-}) + (O^{2-}) = (O^{2-})_e \tag{15}$$

where $(O^{2-})_e$ may be described as an "equivalent oxide concentration."

Eliminating the carbonate concentration in Eq. 15 by substitution from Eq. 14, we have

$$(O^{2-})(K_1p + 1) = (O^{2-})_e \tag{16}$$

which, by substitution into Eq. 13, eliminates the oxide concentration, yielding an expression for the potential difference between two carbon electrodes in terms of their local CO_2 partial pressures:

$$E_2 - E_1 = \frac{RT}{4F} \ln \frac{p_2(K_1p_2 + 1)^2}{p_1(K_1p_1 + 1)^2} \tag{17}$$

For all conditions of interest, K_1 is sufficiently large that K_1p is much greater than unity, hence Eq. 17 may be simplified to the form

$$E_2 - E_1 = -\frac{3RT}{4F} \ln \frac{p_2}{p_1} \tag{18}$$

Under the conditions postulated for this discussion of equilibrium relations, Eq. 18 has been found to be in accord with experimental evidence, and an example of this relationship as experimentally determined is presented in Figure 2.

The anode-to-cathode potential difference of Eq. 11 was written on the basis of common ionic concentrations at the two electrodes. In order to include the effect of the difference in oxide ion concentration between anode and cathode, reflected by a difference in CO_2 partial pressure between these electrodes, we may add the carbon electrode potential difference given by Eq. 18, whence

$$E = 1.025 - \frac{RT}{4F} \ln \frac{p_c^3}{(O_2)_a p_a^2} \tag{19}$$

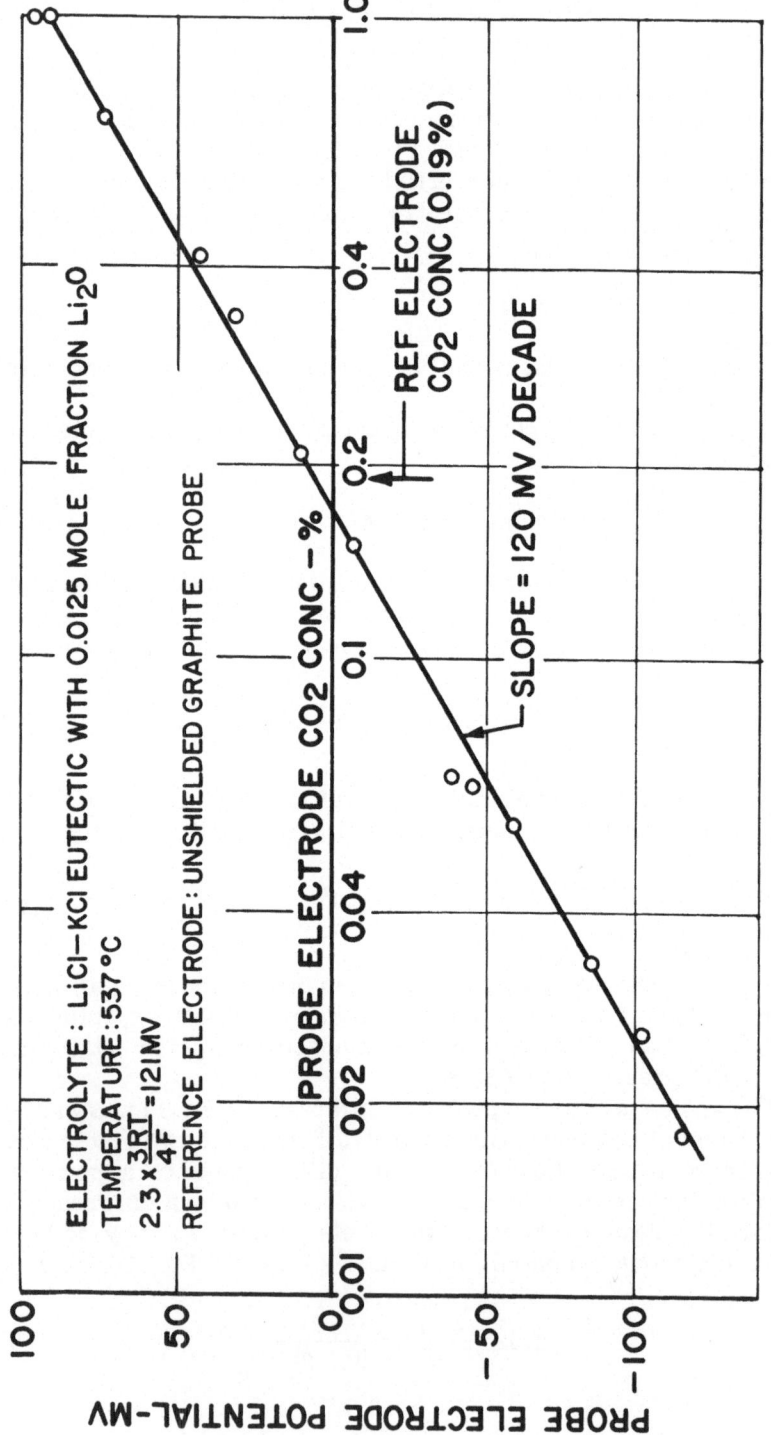

Figure 2.   Carbon electrode potential difference.

in which the subscripts a and c denote anode and cathode quantities, respectively. The CO_2 partial pressure at the cathode during cell operation normally will be several orders of magnitude lower than at the anode, hence the open-circuit cell voltage, measured by establishing steady-state electrolytic conditions and suddenly breaking the circuit, will be appreciably greater than previously indicated for a balanced equilibrium in the melt. For example, letting $p_a = 0.0005$, as before, but taking $p_c = 10^{-6}$, the open-circuit cell voltage will be 1.464 volts, and the limiting efficiency at zero current will be 70%. Successful operation has been achieved with applied potentials in the range 1.6 to 2.0 volts, indicating efficiencies ranging from about 65% to 50% for the over-all chemical/electrochemical process.

CARBON DIOXIDE IN EFFLUENT GASES

As mentioned earlier, effective absorption of CO_2 from a dilute atmospheric gas mixture (space cabin atmosphere) that contacts the electrolyte by bubbling, liquid spraying, or any other gas absorption technique, requires that the equilibrium CO_2 pressure of the melt be no greater than the CO_2 pressure in the effluent gas. Oxygen leaving the anode region will also be mixed with CO_2 at equilibrium pressure for a very slow process (low anode current density), or, more specifically, at the CO_2 partial pressure in the immediate vicinity of the anode, where depletion of oxide ions results in an increase in the ratio of carbonate to oxide ion concentration. Thus, the anode gas output may be enriched in CO_2 above the level obtaining in the bulk of the melt. In the limit, for extremely high anode current density, where ionic diffusion is inadequate to replenish oxide ions, 2 moles of CO_2 will be produced for every mole of oxygen, and the gas mixture will be one-third oxygen and two-thirds carbon dioxide. Since this would impose an unacceptable requirement for recycling anode gas for reabsorption of CO_2, it is obvious that anode current density must be limited to a value that permits mixing processes in the melt (convection, diffusion) to hold a near-equilibrium CO_2 partial pressure in this region.

It is fortunate that the solubility and the diffusion rate of oxygen in eligible electrolytes are sufficiently low that the transport of dissolved oxygen, either from the scrubber or from the anode to the cathode, may be ignored. While solubilities and diffusion rates of various gases in alkali carbonate electrolytes have not been measured directly, laboratory experience has shown that, except for carbon dioxide and water vapor, other dissolved gases, such as oxygen, are not transported to the cathode in appreciable quantities relative to the rate of carbon deposition. It is nevertheless necessary to interpose a shield that prevents gas bubbles from reaching the cathode by convection, since these can directly oxidize the carbon deposit and drastically impair process efficiency. However, with simple shielding, the Faraday current efficiency of the cell, measured relative to either of the desired output materials, carbon and oxygen, is found to be substantially 100%.

CARBON MONOXIDE EQUILIBRIUM

Certain other gases are relevant to the cathodic process and impose significant restrictions on the operating conditions required for adequate performance of the system in application to space cabin oxygen reclamation. First, since elementary carbon is produced at the cathode in an environment that can yield carbon dioxide at a finite pressure, the reaction between carbon and CO_2 to produce carbon monoxide must be recognized. An equilibrium will tend to be established in accordance with the relations

$$C + CO_2 \rightleftharpoons 2CO, \qquad K_3 = (CO)^2/(CO_2) = c^2/p \tag{20}$$

where, for brevity, c represents the partial pressure of carbon monoxide. If, as earlier, we take the working level of p at the cathode to be 10^{-6}, c will be about 1.5×10^{-4}, or 150 p.p.m. in any gas mixture above the electrolyte surface at the cathode. At the least, it would be undesirable for reasons of toxicity to let this concentration of carbon monoxide appear in the cell output; hence the cathode region must be shielded from mixing with the air leaving the cell as well as from bubble communication in the melt. Figure 3 shows the variation of certain equilibrium constants with temperature, obtained from published free energy data, where the curve for K_3 indicates that carbon monoxide would become increasingly prominent as a cathode gas as temperature rises. However, the significance of carbon monoxide lies in another area, rather than in the toxicity and shielding aspect, as will become apparent later.

WATER VAPOR EQUILIBRIUM

Thus far we have considered a process in which the electrolyte comprises lithium carbonate and lithium chloride, with oxide ions produced by dissociation of carbonate in equilibrium with carbon

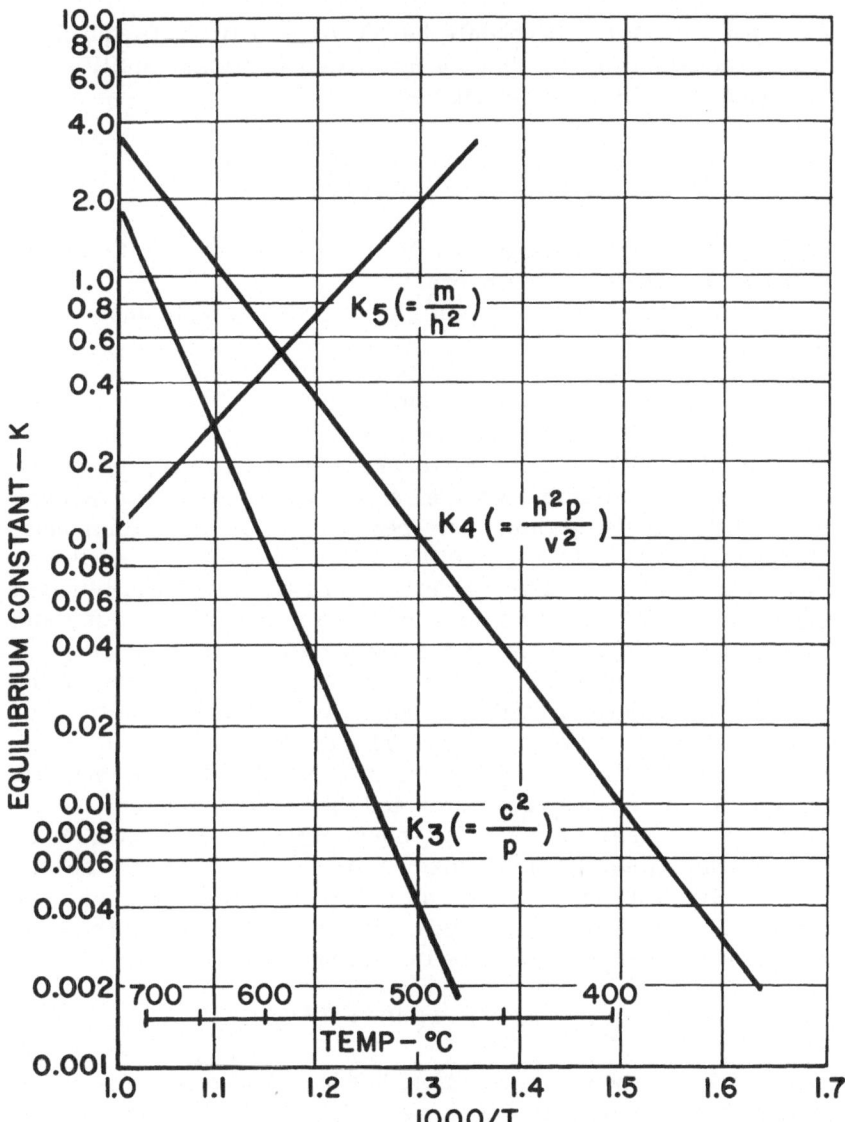

Figure 3. Equilibrium constants.

dioxide. The input gas was presumed to be an oxygen-nitrogen mixture including a small percentage of carbon dioxide, and no mention was made of the possible presence of water vapor. A space cabin atmosphere would inevitably have a significant humidity, probably with a dew point in the neighborhood of 50°F., or about 0.012 atm. partial pressure of water vapor. In scrubbing this gas with an oxide-containing melt, there is not only the CO_2-oxide-carbonate equilibrium to be considered, but also the H_2O-oxide-hydroxide equilibrium, in accordance with the following relations:

$$H_2O + O^{2-} \rightleftharpoons 2OH^-, \qquad K_2 = \frac{(OH^-)^2}{(H_2O)(O^{2-})} = \frac{(OH^-)^2}{v(O^{2-})} \tag{21}$$

The activity of water vapor, (H_2O), is represented by its partial pressure, designated by v for brevity.

The hydrolysis of LiCl might also be considered, in accordance with Eq. 22.

$$LiCl + H_2O \rightleftharpoons LiOH + HCl \tag{22}$$

However, since HCl can immediately react with lithium carbonate so as to replenish lithium chloride, and since no escaping HCl has been detected, even at low parts-per-million levels, with normal water vapor pressure, it is believed that hydrolysis of LiCl may properly be ignored in the consideration of water vapor equilibrium. Hydrolysis of LiCl, due to its extremely hygroscopic nature at room temperature, *is* a problem in the *preparation* of the electrolyte if low hydroxide content is sought, and methods for eliminating moisture are relatively tedious, but in the present case we need not be concerned with hydrolysis. The direct absorption of water vapor by oxide ion, as in Eq. 21, may be taken to represent the primary interaction between water and the electrolyte.

As in the case of the "dry" electrolyte, the total number of electrochemical equivalents of oxide, carbonate, and hydroxide ions will be a fixed quantity, hence, as in Eq. 15, we may write an expression for an "equivalent oxide concentration" in terms of the molal concentrations of the constituent ions.

$$(CO_3^{2-}) + (O^{2-}) + \tfrac{1}{2}(OH^-) = (O^{2-})_e \tag{23}$$

Replacing the carbonate and hydroxide concentrations by the use of the equilibrium relations of Eqs. 14 and 21, we obtain an expression relating oxide ion concentration to the partial pressures p and v in terms of the equilibrium constants and the equivalent oxide concentration, also a fixed quantity.

$$K_1 p(O^{2-}) + (O^{2-}) + \tfrac{1}{2}\sqrt{K_2 v(O^{2-})} = (O^{2-})_e \tag{24}$$

The nature of this relationship is shown in the plot of Figure 4, based upon the solution in the form,

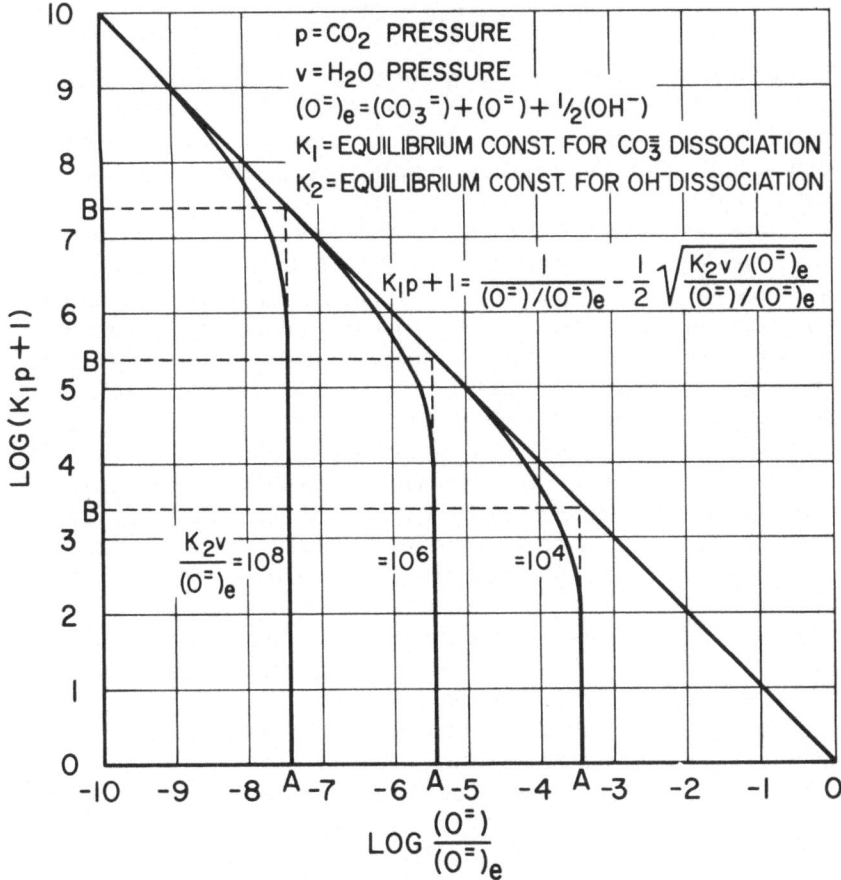

Figure 4. Partial pressures of CO_2 and H_2O in a solution of Li_2CO_3 and LiOH in LiCl.

$$K_1p + 1 = \frac{1}{(O^{2-})/(O^{2-})_e} - \frac{1}{2}\sqrt{\frac{K_2v/(O^{2-})_e}{(O^{2-})/(O^{2-})_e}} \qquad (25)$$

Since K_1p is of interest only for values much larger than unity, this is essentially a plot of log pressure against log oxide concentration, with a dimensionless quantity indicative of water vapor pressure as a parameter.

In Figure 4, the oxide concentrations at the vertical asymptotes of the curves, designated by A in each case, satisfy the requirement, from Eq. 25, that

$$\frac{(O^{2-})}{(O^{2-})_e} = \frac{4(O^{2-})_e}{K_2v} \qquad (26)$$

and the corresponding points, B, on the partial pressure scale, indicating the region of transition in slope, are defined by

$$K_1 p + 1 = K_2 v / 4(0^{2-})_e \tag{27}$$

It is convenient to adopt the convention that, for operation above the transition level indicated by B, the electrolyte is termed a "dry melt," while operation below B would denote a "wet melt," these terms indicating the relative significance of the parameters $K_1 p$ and $K_2 v/(0^{2-})_e$.

Note that, for a dry melt, or relatively low water vapor pressure, oxide ion concentration is inversely proportional to CO_2 pressure. A wet melt, in which water vapor effects are predominant, will have an oxide ion concentration determined by and inversely proportional to water vapor pressure, for relatively low CO_2 pressure. A dry melt will absorb CO_2 by conversion of oxide ion to carbonate, while a wet melt will need to release water vapor (increase the partial pressure) in order to provide oxide ions for CO_2 absorption. This may also be viewed as the "displacement" of water from hydroxyl ions by CO_2 to form carbonate ions. In practice, electrolysis will provide a continuous source of oxide ions and, with the electrolytic current regulated in conformity with the CO_2 absorbed, there will be no variation in either CO_2 or H_2O partial pressure. Nevertheless, for excursions in input CO_2 pressure, the melt CO_2 pressure would be variable in the absence of water vapor but more stable under the conditions of a wet melt.

Absorption of carbon dioxide from a humid atmosphere thus involves a concurrent interaction between water vapor and the absorbent electrolyte. If the process is to provide CO_2 to the melt for electrolytic decomposition, rather than to carry on some alternative process with water vapor, it is necessary that the water vapor input and output be identical. In terms of equilibrium relations, therefore, the water vapor partial pressure of the electrolyte must be identical with the vapor pressure in the processed atmosphere. Thus, it becomes necessary to devise and achieve an electrolyte composition that will satisfy requirements for two partial pressures, p and v, in order that CO_2 be effectively absorbed and that the H_2O content be unaffected by the passage of space cabin air through the cell. Alternatively stated, the hydroxide and carbonate concentrations must satisfy the relationship

$$\frac{K_2}{K_1} = \frac{(OH^-)^2 p}{(CO_3^{2-}) v} \tag{28}$$

The constants K_1 and K_2 vary with temperature; hence, choice of a melt composition for a desired equilibrium restricts operation to a unique temperature. On the other hand, specification of operating temperature, together with the partial pressures p and v, fixes only a concentration *ratio,* and some freedom is thus allowed in arriving at a composition that also has desirable qualities with regard to melting point, ionic conductivity, vapor pressure of constituent salts, and other properties that influence such factors as performance, materials compatibility, and mechanical design.

Gas-forming reactions at the cathode

In the presence of water vapor, other gases in addition to carbon monoxide may be formed at the cathode by chemical interaction between water vapor and carbon. Hydrogen and methane have been detected in measurable concentrations; hence equilibrium relations that specify these gases will be postulated. The actual mechanisms for the formation of hydrogen and methane have not been established; the postulated equilibrium relations that follow are, therefore, primarily symbolic. Nevertheless, whatever the true mechanisms may be, the postulated relations must be quantitatively satisfied in a thermodynamic sense for equilibrium, and thus are significant in the formulation of further restrictions on the process conditions.

Let us consider the reaction of water vapor with carbon to produce hydrogen and carbon dioxide:

$$C + 2H_2O \rightleftharpoons 2H_2 + CO_2, \qquad K_4 = h^2 p / v^2 \tag{29}$$

Here, h represents the partial pressure of hydrogen, which is uniquely determined, at any given temperature, by p and v. As discussed for carbon monoxide, hydrogen at the indicated partial pressure will appear in the gas mixture above the electrolyte at the cathode, and since, for typical operating conditions, this may be a substantial fraction of atmospheric pressure, an additional reason is provided for shielding the cathode and preventing intermixing of these gases with the atmospheric effluent from the cell. Furthermore, purging hydrogen out of this region would be highly undesirable, since more would be generated by further reaction of carbon with water vapor so long as the equilibrium pressure were not maintained, thereby depleting the carbon deposited, restoring carbon dioxide to the electrolyte, and impairing the efficiency of the cell.

Similarly, methane can also be produced, and we choose the direct reaction of carbon with hydrogen arbitrarily to represent the equilibrium requirement:

$$C + 2H_2 \rightleftharpoons CH_4, \qquad K_5 = m / h^2 \tag{30}$$

The partial pressure of methane is here represented by m. Methane, also, can appear in substantial concentration.

On the basis of confirmation by experimental observation,

equilibrium requirements for five gases associated with the electrolyte and its cathodic carbon deposit have been discussed: carbon dioxide, water vapor, carbon monoxide, hydrogen, and methane. Since coexistence of these gases is required at an electrolyte surface, as at the cathode, a total gas pressure, g, may be expressed by

$$g = p + v + c + h + m \qquad (31)$$

Suppose that the total pressure g were to exceed 1 atm., and the cathode region above the electrolyte surface were vented so as to maintain 1 atm. These gases then would be generated continuously, form into bubbles, and rise to the surface and escape, to the extent that they were not restrained by adherence to the electrode surface. The growth of such bubbles could be expected to be deleterious to the orderly formation of a carbon deposit on the cathode. Formation of craters, fissures, and other types of gas passages in the carbon deposit, as has been observed in laboratory experiments, may be ascribed to such a phenomenon. Therefore, a rational criterion for satisfactory process control may be established on the basis that the equilibrium gas mixture total pressure must not be allowed to exceed the total pressure on the melt. Furthermore, since the working process is one of steady-state nonequilibrium, some allowance must be made for local departures from equilibrium at the cathode, setting the equilibrium gas pressure limit somewhat lower than the operating pressure by a suitable margin, empirically determined.

Introducing the representations previously derived for the several gases, in terms of p and v, we have

$$g = p + v + \sqrt{K_3 p} + v\sqrt{K_4 / p} + K_5 K_4 \, v^2 / p \qquad (32)$$

If p and v are determined by the engineering specifications for the function of the cell, setting a limit on g is equivalent to determining a limiting operating temperature, since temperature, which determines K_3, K_4, and K_5, is then implicitly a dependent variable. In view of the relation between carbonate and hydroxyl ion concentrations determined by v/p (Eq. 28), it then remains to determine the concentration of diluent solvent, such as LiCl, necessary to achieve a melting point suitably below the limiting operating temperature.

Figure 5 presents calculated curves of gas partial pressures against reciprocal temperature, based on the equilibrium constants calculated and plotted in Figure 3. The values of v and p were

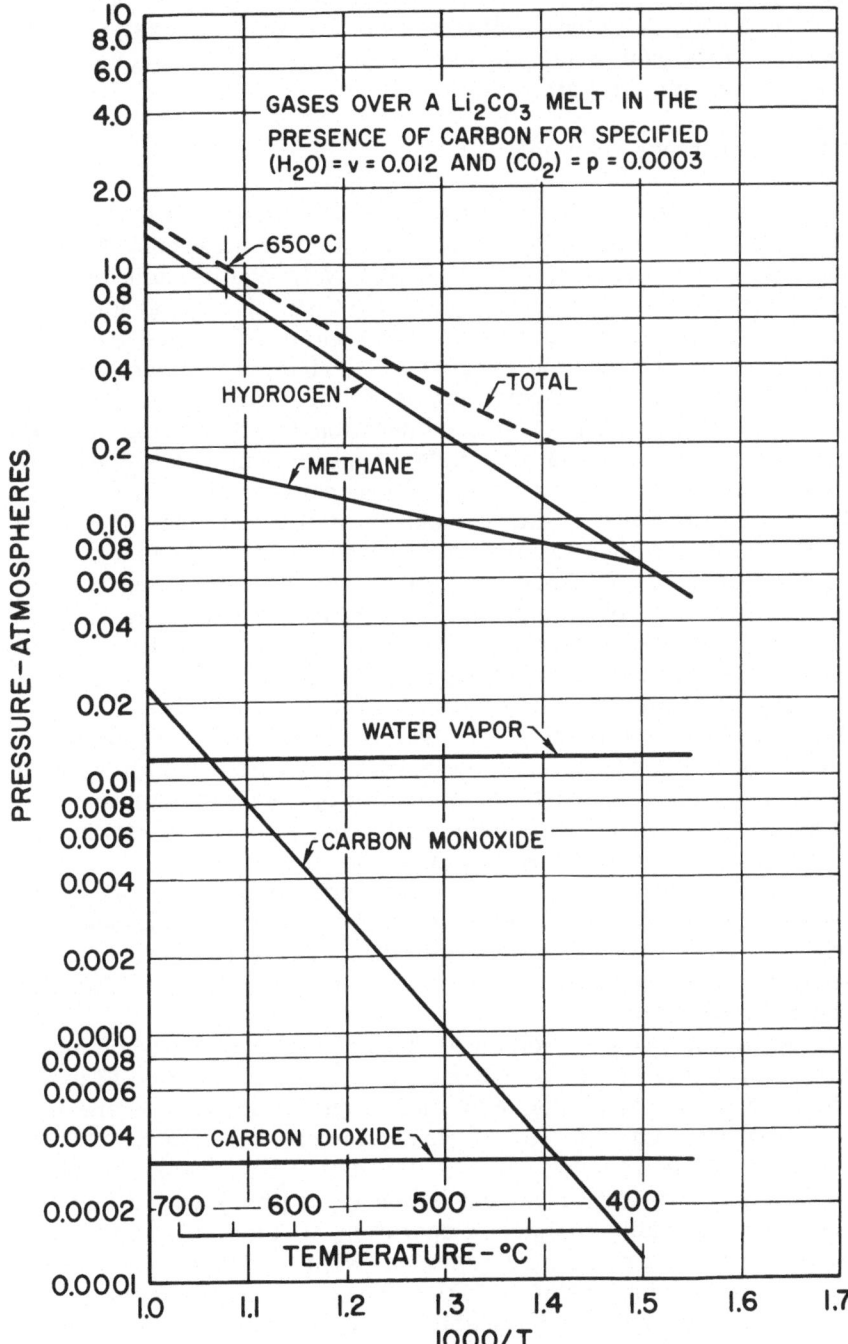

Figure 5. Calculated equilibrium pressures.

chosen arbitrarily and are based on a normal humidity level corresponding to 50°F. dew point and normal atmospheric CO_2 concentration of 0.03%. It is interesting to note that the total gas pressure reaches sea-level atmospheric pressure at 650°C., and at 550°C. the gas pressure is $\frac{1}{2}$ atm.

Carbon electrode potential

When in an earlier section (page 81), we considered the efficiency of the electrolytic cell, the potential difference between two carbon electrodes, each separately in equilibrium with the local electrolyte, was given by Eq. 13. Concentration gradients in the electrolyte do not contribute a potential difference due to ion transference because the current is carried predominantly by the lithium ion, whose concentration is uniform throughout, being the only cation in the case under consideration. By introducing the concept of a fixed equivalent oxide concentration, Eq. 18 was derived, relating the potential difference between two carbon electrodes to the ratio of CO_2 pressures at the two electrodes. It is evident that this suggests the use of a carbon electrode as a sensing device for local CO_2 pressure, and that by controlling the CO_2 pressure at a carbon electrode, a reference potential might be established. This would be useful as a working standard against which other electrode potentials might be independently measured, as, for example, in polarography. Attempts to devise and apply other types of reference electrodes, particularly an oxygen reference, encountered problems in electrode interactions with the melt and consequent electrolyte contamination by undesired ionic species. A carbon electrode, on the other hand, is not a source of contamination, being a normal product of the cathodic process.

Carbon electrodes have been successfully used both as references and as probes, or sensors, in our work, and their behavior appears to be in good accord with theory. However, a further extension of the approach presented earlier is necessary to describe their performance.

With the presence of water vapor in a hydroxide-containing electrolyte, the oxide ion concentration required by Eq. 13 is defined by Eq. 25, requiring solution of a quadratic form before substitution into Eq. 13. The algebraic format tends to obscure the meaning of the result, but since the plot of CO_2 pressure against oxide concentration for various water vapor pressures in Figure 4 indicates separable regimes of operation, it is more illuminating to consider two extremes, represented by predominance of the carbonate-oxide relationship at high p, low v, and by predominance of the hydroxide-oxide relationship at low p, high v. The first of these cases was essentially applied in the derivation of Eq. 18, which is repeated here for convenience:

$$E_2 - E_1 = -\frac{3RT}{4F} \ln \frac{p_2}{p_1} \tag{18}$$

From Eq. 25, when the water vapor pressure is sufficiently high so that oxide concentration is relatively independent of CO_2 pressure, the oxide concentration is found to be given by

$$(O^{2-})/(O^{2-})_e = 4(O^{2-})_e / K_2 v \tag{33}$$

Substituting this expression into Eq. 13 in order to eliminate the oxide concentrations, we obtain

$$E_2 - E_1 = -\frac{RT}{4F} \ln \frac{p_2 v_2^2}{p_1 v_1^2} \tag{34}$$

where the equivalent oxide concentration and the equilibrium constant K_2 do not appear, due to cancellation, being the same at both electrodes.

Equation 18, plotted on semilog coordinates as in Figure 2, becomes a linear relationship between potential difference and CO_2 pressure, with a voltage coefficient, measured by the slope, of $3RT/4F$, or, with logarithms to base ten at 550°C., a coefficient of 124 mv./decade.

Equation 34, if water vapor pressure is identical at the two electrodes, represents a similar relationship but with a different voltage coefficient, $RT/4F$, which would correspond to a slope of 41.3 mv./decade at 550°C. Thus, for a given water vapor pressure, the potential difference between carbon electrodes tends to be linearly proportional to the logarithm of the CO_2 pressure ratio, with the $RT/4F$ slope at low pressures but changing to the $3RT/4F$ slope at higher pressures. This transition in slope has been observed experimentally, and a typical example is shown in Figure 6. It is interesting that the slope is changed by a ratio that is very nearly three, as pressure is varied from below to above a transition region, but it may also be significant that there is a discrepancy in the absolute values of the slopes. Whereas the theory indicates slopes of 41.3 and 124 mv./decade, these experimental data show slopes of 38 and 117 mv./decade. Although the precision of representation of the data by the straight lines may be questionable, the preponderance of evidence indicates a tendency toward lower than theoretical slopes, generally beyond the uncertainty possible through inaccuracies in temperature. Furthermore, the transition region

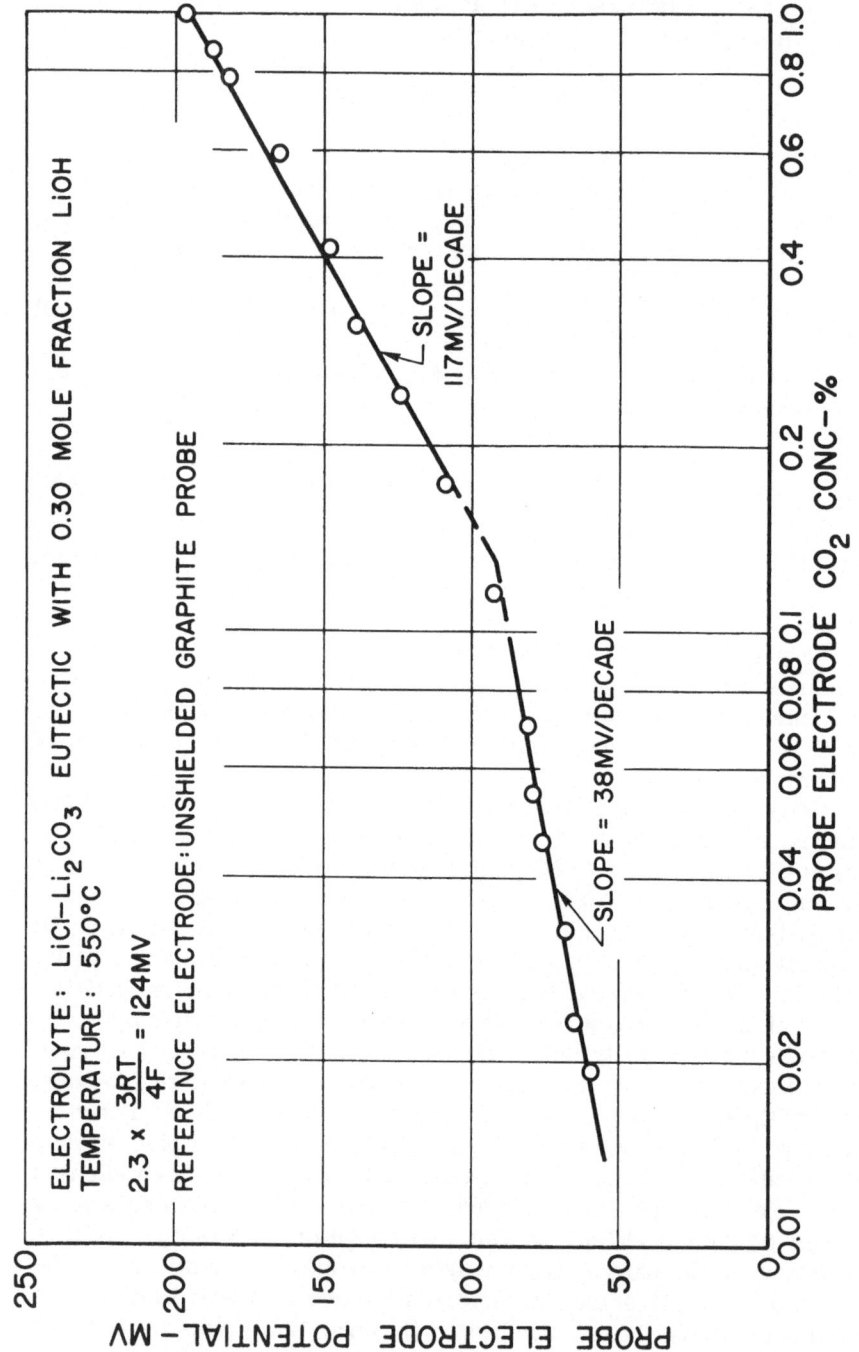

Figure 6. Carbon electrode potential difference.

appears to be more sharply defined than would be inferred from the curves of Figure 4, where the stabilization of oxide ion concentration by water vapor seems to take place in a transition region extending over more than two decades of pressure. We are led to believe that the process and equilibrium relationships on which the prediction of slope is based must be inaccurate or incomplete and that there is some further variable that is not under control. It is likely that the problem lies in the interaction between water vapor and the melt, since the discrepancy in slope appears commonly to be associated with large or inadequately controlled water vapor pressure. Nevertheless, during any particular experiment, the potential:pressure relationship is found to be unique and reproducible; hence it can be useful through calibration.

The electrode configuration used to obtain these measurements, which is therefore also applicable to the function of a reference electrode, is shown in Figure 7. A graphite rod of spectrographic

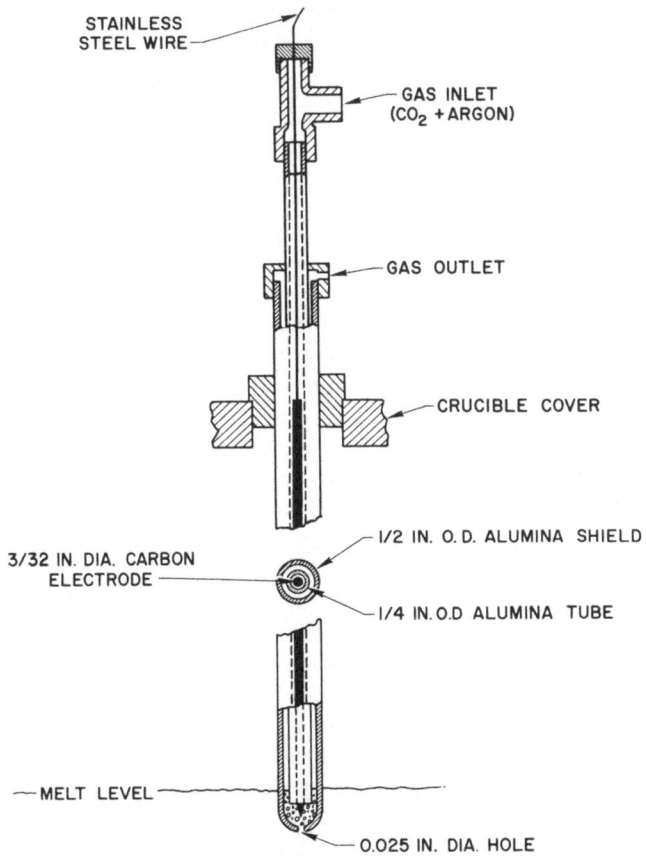

Figure 7. Carbon reference electrode.

quality is inserted into the electrolyte through a close-fitting tube of aluminum oxide, projecting into the electrolyte about two diameters beyond the end of the tube. A controlled humidified mixture of carbon dioxide and argon is fed to the electrode tip through the clearance space inside the tube, to escape over the tip in a fine stream of bubbles. The CO_2 in the gas stream rapidly establishes carbonate-oxide equilibrium in the vicinity of the electrode tip, which is restricted by enclosure within an outer shield, also of aluminum oxide. The shield also serves to isolate the effluent gas stream from the bulk of the electrolyte, restricting the controlled environment to the small volume of electrolyte within the assembly. In order to provide ionic conductivity to the outside melt, the lower end of the shielding tube is punctured by a small hole, effectively constituting a salt bridge to the surrounding electrolyte.

This configuration may be used to provide a reference potential for voltage measurements on other electrodes in the same cell. Also, by using two such assemblies and holding CO_2 partial pressure constant at one while varying CO_2 partial pressure at the other, the characteristics of a carbonate melt of unknown composition can be obtained in the form of an E versus log p curve as in Figure 6, thus to determine the location of the transition region.

MATERIALS COMPATIBILITY

While the subject of corrosion and electrolyte contamination by container and electrode materials is primarily an engineering development problem beyond the scope of the research program discussed here, it seems nevertheless appropriate to mention several aspects of experience accumulated in more than 2 years of laboratory experimentation with fused carbonate electrolytes. Molten mixtures of alkali carbonates and halides are potentially reactive with many common materials, especially in the presence of oxygen and, although chemical theory may provide some guidance, the state of the art rests primarily on laboratory observation.

For limited experimental purposes (as, for example, in polarographic studies, where avoidance of electrolyte contamination by foreign ions is important), graphite crucibles have been outstandingly successful containers for the melt. However, they are useful only when oxygen may be excluded and when the generation of CO, H_2, and CH_4 is of little consequence; hence carbon is not a universally suitable material. Nickel, stainless steel (AISI 304), and gold-palladium alloy (80% Au) may be used as containers for extended periods, but a slow oxidaton occurs that can ultimately result in objectionable contamination. Most satisfactory experience thus far has been with 99% purity alumina crucibles, whose solubility in various electrolyte compositions used appears to be extremely small and, moreover, not deleterious to the electrolytic process. Alumina is preferable to nickel for electrode shields, especially in the presence of oxygen above the melt surface. Magnesia has been suggested on the basis of its successful use by others in the literature, but no direct experience has been acquired in this program.

The choice of an anode material is especially difficult and appears to restrict the choice of melt composition. All of the metals mentioned suffer dissolution when used as anodes in an electrolyte containing a high mole fraction of chloride (95 to 99%), but the problem is significantly ameliorated in the vicinity of the eutectic mixture of Li_2CO_3 and LiCl (70% mole fraction of LiCl). Nickel has been satisfactorily used as an anode in a mixture based on the eutectic but with LiOH added to reach equilibrium with water vapor at moderate dew points. Under these circumstances, a protective oxide coating of reasonable stability appears to form on the anode. It is expected that a combination of chemical preconditioning and the use of an optimum anode current density can be devised to achieve a passivated oxide coating, but no firm conclusions can yet be drawn.

The cathode material is more specifically a design and development problem, since the electrode surface becomes carbon shortly after electrolysis is started. Most favorable adherence is obtained on iron, stainless steel, nickel, and cobalt, but the choice of substrate for an oxygen reclamation system will undoubtedly depend more on factors such as cathode configuration and carbon removal technique.

COMPOSITION OF ELECTROLYTE

The basic process concept introduced at the outset, whereby lithium is electrolytically reduced to provide an oxidizable reactant for the decomposition of carbon dioxide, would superficially appear to be applicable to other alkali metal carbonates. Calculations of equilibrium compositions of a number of alternatively conceivable systems on the basis of the thermodynamic properties of their reactants and products, with the restriction that such systems must be physically possible, in accordance with the Gibbs phase rule, indicate that the alternatives are severely limited. This work, which is incomplete, will not be discussed here in detail, but the results with regard to lithium, sodium, and potassium salts will be mentioned, since significant constraints are indicated on the choice of electrolyte composition.

Experimental attempts to deposit carbon at the cathode in

melts based on sodium and potassium carbonates, both separately and in combination, as well as in combination with lithium carbonate, have met with varying degrees of success. Calculations of thermodynamic equilibrium in closed systems with arbitrarily fixed quantities of carbon, oxygen, and alkali metal atoms have provided a tentative framework for the evaluation of these experiments, and the general result has been to indicate that the presence of lithium ions at significant concentration appears essential to a successful process of the type under consideration.

According to the results of the calculations, there are only two product combinations in which solid carbon can exist, $M_2O—C—M_2CO_3$ and $M—C—M_2CO_3$, representing condensed phases in the melt, with M as the alkali metal. The former combination is possible with lithium or sodium as the alkali metal, and the sodium system is restricted to temperatures below $900°K$. The latter combination is possible only with sodium or potassium. Coexistence of carbon and an alkali metal in appreciable concentration at the cathode is undesirable for two reasons: first, a significantly higher potential is required for cell operation when the alkali metal is present in finite quantity at the cathode; second, sodium and potassium metals tend to form interstitial compounds with carbon that are physically unstable, resulting, for the case of potassium, in relatively violent disintegration of the deposit. However, although the dissolution of the cathodic deposit has been observed experimentally in electrolytes where sodium and potassium compounds were predominant, an apparently stable deposit has been achieved in a lithium-potassium melt. Calculations have not been made for a two-metal system, but we infer that, since the lithium system is based on electrolytic reduction of lithium at a lower potential than would be required for potassium, no appreciable amount of potassium will be reduced in the two-metal system, and only the $Li_2O—C—Li_2CO_3$ relationship will be pertinent.

On the other hand, although the same type of reasoning would indicate that a sodium-potassium system, operated in a suitable temperature range, should be acceptable, experimental results appear to indicate carbon deposits of unsatisfactory quality. Similarly, a three-metal system, specifically, the ternary eutectic mixture of lithium, sodium, and potassium carbonates, despite its favorable melting point ($390°C.$), has shown inferior performance. At the present state of knowledge, therefore, it appears that lithium salts are essential, potassium salts may be tolerable as additives, but that the inclusion of sodium salts is of questionable value and possibly detrimental.

Although lithium carbonate, lithium oxide, and lithium hydroxide are the only essential ingredients of the electrolyte insofar as requirements for chemical process and gaseous equilibrium are concerned, the relatively high melting point of lithium carbonate (726°C.) makes it necessary to include suitable additives or a solvent so as to permit operation at the lower temperatures indicated by the necessity for inhibiting gaseous outputs at the cathode. As previously mentioned, the use of lithium chloride as a solvent is a conservative possibility, since the eutectic mixture melts at 509°C. A ternary mixture of lithium carbonate, chloride, and fluoride is also suitable, and potassium chloride or fluoride may be incorporated without deleterious effects. From an engineering point of view, it would be desirable to operate the cell at as low a temperature as possible. No limiting low temperature has yet been deduced on the basis of the chemical or electrochemical requirements of the system, so that the problem of reducing the temperature is primarily one of discovering a combination of the eligible ingredients that will achieve a low melting point, have acceptable conductivity and viscosity, and provide adequate solubility for the increased oxide concentration developed at the cathode during steady-state nonequilibrium operation.

DISCUSSION

In the preceding sections a number of constraints have been described for the basic process concept of carbon dioxide decomposition by chemical reaction with an electrochemically reduced alkali metal. The fundamental process is deceptively simple; it will take place as described and satisfy the functional requirements for engineering application only if adequate control is exercised over the internal and external environmental factors that determine the course of the reactions required. In summary, the following conditions must be imposed:

1. Carbonate-oxide-hydroxide concentration balance in the electrolyte must establish suitable CO$_2$ and H$_2$O concentrations such that CO$_2$ is absorbed at low partial pressure and H$_2$O is not.
2. Electrolyte composition must permit an operating temperature such that the total pressure of gases required for equilibrium at the cathode does not exceed ambient pressure.
3. Contamination of the electrolyte by materials that may alter the electrochemical or chemical processes at the electrodes must be avoided.

Some further comment on the third condition is appropriate. Dissolution of the melt container, thermocouples, cathode shield, anode, or any other system elements in contact with the electrolyte can introduce into the melt foreign ions that may be reduced and contribute outputs at the cathode and/or at the anode if their redox potentials are lower than the normal operating potentials. Nickel, iron, and cobalt are prime offenders in this regard, resulting in significant contamination of the carbon deposit at trace contamination levels and in virtual exclusion of the carbon-forming process at higher levels. This also results in a loss in capacity for absorbing CO_2, since the oxide ions formed in the cathode process are required to maintain the ability to absorb.

The general quality of the cathodic carbon deposit is a subject that cannot be discussed conclusively at this time, this being the primary area of concern in the continuing program of research. The objective of 100% purity and high density has been approached, but continuous accomplishment of this goal requires better control of the cathode environment and, probably, more detailed understanding of the cathode process than has yet been achieved. Nevertheless, the formation of a porous, cratered, fissured deposit showing evidence of gas evolution during deposition, commonly experienced in early experiments, is avoidable by compliance with the requirements here discussed for equilibrium of gases at the cathode. A relatively smooth-surfaced carbon deposit, without gas-containing voids, can be produced repeatedly.

However, there is a tendency for entrapment of electrolyte salts in the microstructure of the carbon deposit, possibly because of the dendritic nature of the deposit formed, which tends to inhibit the diffusion of oxide ions away from the sites of cathodic reaction. The carbon purity thus is found to approach 100% in the initial layers but decreases due to occluded salts beyond the first several thousandths of an inch, so that the average composition in a deposit of several millimeters thickness may be as low as 50% carbon, this figure being a strong function of current density.

From an engineering point of view, less than 100% carbon in the deposit does not constitute an insurmountable problem, since the extraction of the deposit, particularly in zero gravity conditions, represents a demand for design ingenuity in any case. Since the occluded salts are soluble in the melt and presumed to be liquid in the deposit as formed, carbon removal by expressing salts through a filter and formation of a carbon cake appears to be a feasible process.

The need to devise a system configuration suitable for operation in zero gravity poses engineering design problems in a number of areas, since the electrolytic cell concept inherently involves interfaces between liquid and gaseous phases as well as between liquid and solid phases. It is evident that a rotating, axially symmetric cell could provide a centrifugal field as a substitute for gravity, but more elegant approaches are under study to avoid the necessity for slip rings, rotating seals, and structural compromises. Possibilities in the use of surface tension, vortex flow, porous matrices, and other principles for liquid stabilization or containment in zero gravity, however, are beyond the scope of this research study.

The primary features of the electrolytic system for carbon dioxide decomposition that make it attractive for oxygen reclamation in a space vehicle are its inherently high efficiency ($\sim$50%) and its ability to exchange oxygen for carbon dioxide in the processed gas stream without the need for previous separation or concentration of carbon dioxide.

5.

CARBON DIOXIDE CONVERSION
FOR OXYGEN RECOVERY

JOHN F. FOSTER

Battelle Memorial Institute, Columbus, Ohio

The requirements for maintaining a suitable atmosphere for human occupancy of a space cabin are as complex and as interrelated as the physiological processes that maintain the health and well-being of the occupants themselves. The major need is a continuous supply of breathable oxygen to balance consumption. This is perhaps the most difficult and immediate problem as goals are set for longer missions that dictate the regeneration of consumed materials from wastes.

Of the three major life-support categories—water, food, and air—the renewal of the oxygen in the cabin atmosphere represents a more urgent problem than do either water or food supplies, for two reasons: The water supply can be renewed with relative ease, using a closed water cycle, because water is a chemically stable compound that may be contaminated but is not altered. Techniques for water purification are well known, and equipment adapted to space use is already developed as engineering prototypes for use in missions of perhaps two or more weeks. On the other hand, a completely closed food cycle is beyond the capability of present space technology. Studies pointed toward producing some food products from wastes for partial closure of the food cycle are in their early stages, and actual mission applications are probably some years away. Nevertheless, stored supplies can satisfy the major part of food requirements for periods of perhaps a year or more, so long as sufficient pure water is available from the closed water cycle to reconstitute dried and concentrated food products. Thus the critical remaining need is for carbon dioxide conversion to extend immediate mission capabilities to much longer periods than the 30 or 60 days now possible with one or another form of stored oxygen supply.

This discussion compares a number of different concepts for the conversion of CO_2, using a qualitative appraisal of their relative complexity and reliability as estimated by specific criteria described later. Also considered are the engineering difficulties that may be met in combining the CO_2 conversion with other parts of an integrated oxygen-recovery cycle. In general, no attempt is made to summarize the available quantitative calculations of weight and power requirements, but some references are cited on recent studies that have furnished such information. These have been helpful in considering the over-all picture of carbon dioxide conversion. Most attention is given in the following discussions to chemical methods for CO_2 conversion, but a brief interpretation of the relative importance of chemical methods versus biological processes is included. Biological processes combine CO_2 conversion for oxygen recovery with the production of by-products, which are a potential addition to the total food supply.

CRITERIA FOR CHEMICAL CONVERSIONS

The four chemical processes that occupy the major part of this discussion survived a preliminary screening according to at least five criteria that were applied to predict their practical usefulness. The processes, namely, the Sabatier reaction, the Bosch reaction, electrolysis in a solid electrolyte, and electrolysis in a molten-salt electrolyte, all have favorable characteristics with respect to one or more of the following criteria: (1) separability of products; (2) operability at zero gravity; (3) low stresses on materials and components; (4) potentially low toxicity; and (5) minimum number of process steps.

A superior rating by some of the foregoing criteria indicates qualitatively favorable weight and power requirements, which justify research and engineering development to overcome the deficiencies of the over-all process as judged by the remaining criteria.

The listed criteria incorporate certain special meanings of the terms used, which are defined here to clarify the later descriptions of individual systems. Each chemical conversion of carbon dioxide is based on a concept of its chemical reaction with some other material. The reaction may be known; hypothetical; or even a model reaction that is the over-all sum of a number of intermediate steps that may not be completely understood. In any event, this reaction defines the concept and specifies the products that are obtained. The separability of these products is appraised according to the simplicity or efficiency of the treatment by which they may be isolated from each other. Commonly, one product carries the oxygen content of the carbon dioxide, and it must be treated in a subsequent step to release free oxygen. The other product may also contain part of the oxygen as well as the carbon. It may be recycled, discarded, or used as feed to additional steps in the total oxygen-recovery system.

Operability at zero gravity implies material transfer to feed the reactor and through the separation processes, without the necessity for separation of discrete phases by relying on differences in density. Ordinarily this requirement may be met in a number of ways, such as fluid flow down a pressure gradient, or the use of capillarity to transfer a liquid phase by surface tension. The requirement may be obviated if the system is to be used in a natural gravity field. Presumably most systems can be subjected to an artificial gravity field by using centrifugal force, with proper design of the equipment for this purpose. However, the conversion system is apt to be much more versatile if it is gravity insensitive, since the mission is usually not precisely defined during the period of development.

In considering the desirability of low stresses on materials and components, the term "stress" is used in a sense that includes more than mechanical stress; it implies the absence of temperature extremes, the use of low or moderate flow rates in material transfer, mild corrosion and abrasion tendencies, moderate operating pressures, and the like. A relatively wide range of optimum operating conditions would minimize the requirements for precise controls. This characteristic would also qualify as a low stress.

A minimum number of steps is an advantage. This criterion is applicable because a special feed, such as pure CO_2, or further processing of intermediate products to recover breathable oxygen, is necessary in some systems and not in others. Fewer steps would usually indicate less equipment, fewer maintenance problems, and probably higher over-all efficiency. This is not necessarily so, because the individual steps may be more complex or may require greater stresses on material and components.

HYDROGENATION OF CARBON DIOXIDE

Sabatier reaction

The hydrogenation of CO_2 produces water as a primary intermediate product. Water is separated from the reaction mixture and is fed to an electrolytic cell to recover oxygen and hydrogen as co-products. Oxygen returns to the occupied cabin for maintaining the cabin atmosphere, and co-product hydrogen returns to the hydrogenation process. The feed to the hydrogenation process must be essentially pure CO_2 and hydrogen. Carbon dioxide must be absorbed from the cabin atmosphere at low concentration and released for use by regenerating the CO_2 absorber. Therefore, the hydrogenation reaction is one of three essential process steps in closing the oxygen cycle: collection of CO_2; hydrogenation; and water electrolysis. Reduction of CO_2 by hydrogenation was one of the first chemical conversions investigated, primarily because the electrolysis of water was a relatively well-known and simple process for recovering pure oxygen.

The reduction of CO_2 by hydrogen can occur at three different levels, represented by the hypothetical stepwise reactions 1, 2, and 3, the sum of which is reaction 4:

$$CO_2 + H_2 \rightleftharpoons CO + H_2O \tag{1}$$

$$CO + H_2 \rightleftharpoons C + H_2O \tag{2}$$

$$C + 2H_2 \rightleftharpoons CH_4 \tag{3}$$

$$\overline{CO_2 + 4H_2 \rightleftharpoons CH_4 + 2H_2O} \tag{4}$$

These reactions may occur simultaneously. It is a basic principle that, in any series of such stepwise reactions, the system may be controlled, by using appropriate reaction conditions and catalysts, to favor the predominance of either the first or the last step as a matter of choice, to the practical exclusion of the intermediate steps in the series. The Sabatier reaction favors the last step and consumes the intermediate products so that reduction is essentially completed to give methane and water, as shown in Eq. 4.

Rousseau (1963) has summarized the conditions under which the Sabatier reaction occurs. On the basis of available information he estimated that conversion efficiencies of about 95% could be obtained at a temperature of 600°F. and at atmospheric pressure, with a catalyst consisting of nickel deposited on kieselguhr. The conversion efficiency decreased at reduced pressures.

Recent studies (Thompson, 1964, 1965; Remus et al., 1965) have been directed toward improving catalyst activity so that complete conversion can be reached, the operating temperature can be reduced, and the sensitivity of the catalyst to poisoning can be reduced. Remus (Remus et al., 1965) reports that a catalyst of pure ruthenium metal powder provided 99% conversion of CO_2 at 357°F. and at 1 atm., using a hydrogen-to-carbon dioxide ratio of 4.4 (10% excess hydrogen) and a space velocity of 310/hr. Intermittent exposure of the catalyst to hydrogen sulfide gas did not affect its performance. Operation at reduced pressures down to 5 p.s.i.a. decreased conversion by only 1 or 2%. It thus appears that almost the ultimate in CO_2 conversion has been reached for the Sabatier reaction under these conditions.

According to these data the Sabatier process qualified as promising by several of the criteria cited, and justifies the amount of attention it has already received in research programs. Only two major products, water vapor and methane, are produced. The water vapor as a primary product is easily separated from the methane and unreacted hydrogen by condensation at a suitably low temperature. The process is gravity insensitive, since the catalyst can be held in place by a filter or screen to prevent movement with the gas stream, and the condensed liquid-phase water can be collected, for example, by passing the gas through an absorbent sponge. The temperature and pressure conditions are mild compared to alternate processes, and the selected catalyst is critically sensitive neither to sulfur poisoning nor to changes in pressure and temperature over quite a range. The only potentially toxic by-product is carbon monoxide, and its formation is effectively inhibited by using the stated excess of hydrogen. Formation of carbon deposits on the surface of the catalyst is no problem for the same reason.

The three process steps of CO_2 collection, CO_2 conversion, and water electrolysis are minimal, if the mission requirements permit discard of the methane-hydrogen mixture, which represents about 55% of the hydrogen obtained from the electrolysis step. The discarded hydrogen might be replaced by electrolysis of stored water, or partially from the electrolysis of excess water produced in the human metabolic process of converting food to energy.

The hydrogen must be recovered and recycled for missions of a year or more, because for this length of time any known system involving stored supplies carries an intolerable weight penalty. Then an additional step is needed, which would use a separate system for converting methane by a reaction such as one of the following:

$$CH_4 + CO_2 \rightleftharpoons C + 2H_2O \tag{5}$$

$$CH_4 \rightleftharpoons C + 2H_2 \tag{6}$$

$$nCH_4 \rightleftharpoons C_nH_{m \leq n} + (2n - m/2)H_2 \tag{7}$$

Rousseau (1964) has discussed a number of secondary reactions for the utilization of the methane product from the Sabatier hydrogenation process, including reactions 5 and 6. His plot of the thermodynamic calculations for reaction 5 shows that the reaction equilibrium is displaced toward the right at low temperatures, and the formation of carbon and water is favored. About 80% conversion of carbon dioxide and methane is indicated at 200°F. Conversion decreases to 50% at 1,000°F. and to about 40% at 2,000°F. The optimum temperature and catalyst for this conversion has not yet been established. In any event it will probably be necessary, if this reaction is used, to recycle unconverted carbon dioxide and methane after the removal of water vapor and carbon from the reaction mixture. Water would be condensed by cooling the mixture, and returned to the water electrolysis subsystem for recovery of oxygen. Solid carbon would probably collect on the surface of the catalyst and reduce its effectiveness. The problems of removing carbon from this and other systems where it is a product are discussed later.

As an alternative, the essentially pure methane from the Sabatier reaction might be fed to a pyrolysis reactor where it would undergo reaction 6, with the formation of carbon and hydrogen. Thermodynamic calculations of the equilibrium show that complete pyrolytic conversion to carbon and hydrogen is attained only at temperatures approaching 2,000°F., and that about 50% conversion is reached at 1,000°F. Once again, carbon would probably separate as a solid phase on the surface of the catalyst with the accompanying problem of removing it and maintaining catalyst activity. The formation of carbon as a pyrolysis product is favored by low pressure. Low system pressure requires a reactor of larger volume, or higher flow velocities. These requirements cancel advantages by other criteria because of greater equipment volume and weight.

Reaction 7 is included as a hypothetical model reaction to indicate the possibility that methane might be converted to another hydrocarbon with a lower hydrogen content, which would recover some of the methane hydrogen for recycling. The hydrogen content of the hypothetical product should make it volatile. It would carry through the reactor and could be condensed by moderate cooling for separation from the hydrogen, without fouling the catalyst in the converter with carbon deposits. There is no real counterpart to this reaction that is known to be suitable for space application, considering the weight and power penalties required for this extra step.   However, such reactions have been proposed by many investigators because of the large amount of experience available from industrial research on the conversion of the methane in natural gas to useful high-boiling products. If the hydrogen content of the discarded hydrocarbon product were low enough, it might be possible to compensate for the hydrogen loss by electrolysis of the excess metabolic water available in the closed atmospheric recovery cycle. The Wulff process for the conversion of methane to acetylene is one embodiment of this general idea, but this process uses an electric arc with such a low energy efficiency that it could not be considered for a space application without major improvement, which seems unlikely.

The general conclusion concerning the Sabatier reaction is that long missions will require one of the extra steps discussed earlier to convert methane. The present status of the results from research studies probably favors methane pyrolysis, with recovery of all of the hydrogen and the discard of solid carbon. This combination of steps then becomes the chemical equivalent of the Bosch reaction discussed in the next section.

Bosch reaction

The Bosch reaction is the sum of the two steps shown as reactions 1 and 2, as follows:

$$CO_2 + 2H_2 \rightleftharpoons C + 2H_2O \qquad (8)$$

Some of the earliest work specifically directed toward appraising the usefulness of this reaction for CO_2 conversion and oxygen recovery was begun at Battelle in 1959 (Foster and McNulty, 1961), and was based on a qualitative screening of the advantages of the Bosch reaction as described earlier. Some of the concepts for direct electrolysis of carbon dioxide, without the water electrolysis step, were not known and were not considered in the initial selection of the Bosch reaction. After the conclusion of the experimental feasibility study, the development of the process was continued by others, notably Remus *et al.* (1963). The integrated Bosch system is probably the farthest advanced toward practical application of any of the candidate processes for CO_2 conversion. It combines the CO_2 conversion reactions in one reactor (reaction 8) with the net result of removing only liquid-phase water by condensation and absorption, and solid carbon by adhesion to solid surfaces within the reactor or by filtration of the gas stream. The primary separation of these products is relatively easy, and is gravity insensitive. Secondary removal of the carbon products from the reactor is still a problem for which the best method has not been selected. The formation of gaseous by-products carbon monoxide and methane is inescapable. The solution has been to recycle these gases back to the inlet of the Bosch reactor, where more carbon dioxide and hydrogen are added. A steady state can be established in which the by-products as well as the primary feed combine in the reactor, in some manner not definitely established, to form the separable primary products.

The maximum temperature of the reactor is about 1,300°F., which does not impose any unusual stress on properly selected materials of construction. The requirement for recycling a comparatively large proportion of the total gas through-put is unattractive because of the correspondingly large pumping system, and because the gas stream must be cooled after each pass to condense the water product, and then must be reheated to reaction temperature by heat exchange or power input before re-entering the reactor. An iron catalyst is used. Catalyst composition is not critical, but the iron apparently must be activated in a reducing atmosphere (perhaps hot hydrogen) before use. The catalyst has taken various forms in different research studies, including steel shot, iron plates, screens, and highly dispersed (fluffed) commercial grade steel wool. There is no definite evidence that the catalyst experiences a significant loss in activity during operation because of the carbon deposits or because of poisoning from traces of contaminating materials. There is some evidence from work not related to space applications that the real catalyst may be an iron compound, which may be transported from the iron-carbon interface to the carbon-gas interface during the carbon deposition process, and thus may remain exposed and reactive, even with a high ratio of carbon to iron in the solid deposits. With a high ratio, the small amount of catalyst required might be considered expendable, without a serious weight penalty being involved. No reports have been found of tests with recycling and reuse of some of the iron-carbon mixture as a catalyst, although presumably it might serve as a suitable catalyst for the Bosch reaction.

The carbon monoxide in the gas-recycle loop is potentially toxic, but it should be possible to avoid significant leakage. The Bosch process represents the irreducible minimum number of three steps for complete oxygen recovery by hydrogenation, involving CO$_2$ absorption and regeneration as a pure feed, the Bosch reaction, and electrolysis of the product water.

ELECTROLYSIS OF CARBON DIOXIDE

Solid electrolyte

Chandler and Oser (1962) reported on the feasibility of the direct electrolysis of carbon dioxide in a solid electrolyte consisting of mixed oxides maintained at an elevated temperature of about 1,800°F. Ionic conduction occurs at this temperature under an imposed potential gradient. The solid electrolyte is fabricated as a ceramic tube with a gas-permeable metal electrode on the outer wall and a similar electrode on the inner wall.

Reactions 9 and 10 represent, respectively, the hypothetical thermal decomposition of carbon dioxide to carbon monoxide and atomic oxygen, and the formation of oxygen ions at the cathode,

$$CO_2 \rightleftharpoons CO + O^{\cdot} \tag{9}$$

$$O^{\cdot} \xrightarrow[+2e]{\text{cathode}} O^{2-} \text{ (solid electrolyte)} \xrightarrow[-2e]{\text{anode}} \tfrac{1}{2}O_2 \tag{10}$$

followed by separation of the oxygen by ionic transfer and its release at the anode within the central passage of the tube. The integrated system consists of three steps, in which (1) CO_2 is absorbed from the cabin atmosphere and regenerated to give a pure
CO_2 feed to the reactor, (2) the CO_2 is electrolyzed according to
reactions 9 and 10, and (3) the carbon monoxide co-product is converted by reaction 11

$$2CO \rightleftharpoons CO_2 + C \tag{11}$$

to precipitate solid carbon and recover carbon dioxide for recycle
to the second step.

The solid-electrolyte system has potential advantages and some
disadvantages that are not yet as fully proved by engineering studies
as are the operating characteristics of the Bosch system. The
separability of products is good, because in this electrolytic separation of oxygen no other ionic species seems possible and ionic
movement is unidirectional. The solid carbon separates spontaneously from the CO-CO_2 stream and deposits on the walls of the
secondary reactor or on the catalyst used in reaction 11. Operability at zero gravity is excellent, because no liquid phases are
necessarily present in any of the steps.

Operating temperatures and pressures do not place excessive
stress on the materials and components, since the ceramic-type
solid electrolyte presumably would be chemically stable at operating temperatures, if the carbon monoxide concentration in contact
with the cathode is kept below the level at which the ceramic oxides
might be reduced.

The characteristic brittleness of ceramic materials may require special precautions against brittle fracture due to vibration
or shock, and against cracking caused by dimensional changes in
heating and cooling.

The rate of gas circulation through the recycle loop is probably less than for a comparable Bosch system, because a substantial conversion of carbon monoxide to CO_2 and carbon should
be attainable in each pass through the converter. The converter
operates at a temperature of about 900°F., so that cooling and re-
heating to electrolysis temperature requires less heat exchange
area than does the condensation of water at about 55°F. in the
Bosch system. The carbon monoxide formed is potentially toxic

but the problem of containing it in the recycle loop is no more serious than the same problem in the Bosch system.

It has been suggested as an advantage of the solid-electrolyte system that an electrolytic process can be operated at several times its optimum design capacity with little sacrifice in efficiency, by increasing the operating voltage and thus the rate of separation of the oxygen ions that form the product oxygen. This might be an emergency mode of operation used only for a brief period. Some provision might have to be made for extra recycling capacity to balance oxygen production with a higher rate of reconversion of carbon monoxide to carbon and CO_2.

A similar increase in capacity could also be obtained in the Bosch system by operating the water electrolysis cell at a higher rate, using a reserve supply of stored water, and providing more capacity in the Bosch reactor for converting hydrogen. In the case of the Bosch system only, there is an additional opportunity for an emergency-mode balance by the use of the Sabatier reaction as an alternate to the Bosch reaction during emergency operation. This could be done with a relatively small amount of additional apparatus, using many of the same controls and material-transfer components that are a part of the Bosch system. Thus, the Bosch system may have some advantage over the solid electrolyte system, when an emergency dictates the production of larger than normal amounts of breathable oxygen for short periods of time.

Molten carbonate electrolyte

Chandler and Oser (1962) proposed in 1960 the concept of recovering oxygen from carbon dioxide by electrolysis of molten carbonates. This concept has evolved further and has been studied experimentally until its technical feasibility appears promising. The present concept is equivalent to the direct electrolysis of carbon dioxide in a single reactor that absorbs carbon dioxide from the cabin atmosphere, separates it into its elemental components, accumulates the solid carbon as a waste product, and returns the oxygen in breathable form to the same atmosphere.

In more detail, the process is a complex combination of anodic, cathodic, and liquid-phase chemical reactions represented by the hypothetical series of reactions 12 to 16.

$$Li_2CO_3 \xrightarrow{\text{melt}} 2Li^+ + CO_3^{2-} \tag{12}$$

$$Li^+ \xrightarrow[+e]{\text{cathode}} Li \tag{13}$$

$$4Li + CO_2 \xrightarrow{\text{melt}} 2Li_2O + C \tag{14}$$

$$Li_2O + CO_2 \xrightarrow{\text{melt}} Li_2CO_3 \tag{15}$$

$$2CO_3^{2-} \xrightarrow[-2e]{\text{anode}} 2CO_2 + O_2 \tag{16}$$

These, or an equivalent series of possibly greater complexity, occur at various regions in a molten lithium carbonate-lithium chloride mixture. The lithium chloride is added to form a eutectic melt and thus to lower the operating temperature. The molten lithium carbonate is the active reaction and transfer medium. In the melted state lithium carbonate ionizes to form lithium ions and carbonate ions. The attainment of the net desired reaction depends on the presence of these two ions as well as an inventory of five other materials in the appropriate concentrations and at appropriate locations in the reaction mix. The other five materials are: lithium metal and lithium oxide as intermediates; carbon dioxide as both feed and intermediate; and carbon and oxygen as products.

The reactions occur simultaneously or consecutively as follows: lithium carbonate is ionized to lithium ion and carbonate ion (Eq. 12). By electrolysis, lithium metal is discharged at the cathode (Eq. 13), lithium metal reacts with carbon dioxide gas in contact with the melt to form lithium oxide and carbon (Eq. 14). Carbonate ion is discharged electrolytically at the cathode to form carbon dioxide and oxygen (Eq. 16). The carbon dioxide released at the cathode is immediately reabsorbed by contact with lithium oxide in the melt to regenerate lithium carbonate as shown in Eq. 15. Pure oxygen is released as a gas to the atmospheric gas phase passing through the melt.

By the same criteria used for qualitative evaluation of the other processes, this process shows considerable merit. The separability of the primary products carbon and oxygen is excellent. The process is, however, gravity sensitive, because the combination of liquid and gas phases is a necessity. Therefore, a centrifugal force field will have to be provided by rotating the reactor to maintain electrolyte contact with the electrodes. This would be a relatively difficult engineering problem for this system, but it has been solved for water electrolysis where the operating conditions are much milder. The stresses on materials and components are relatively severe and highly complex in many respects. Any system containing fused salts operates at a relatively high temperature, and construction materials and components are subject to severe corrosion stress. According to the present concept, the whole cabin atmosphere is treated to absorb the atmospheric carbon dioxide directly in the bath without an absorption-desorption step to concentrate the CO$_2$ before this reaction. This entails the use of relatively large and highly efficient heat exchangers.

A balanced reaction system requires controlled circulation and convection of the melt to transfer lithium oxide formed from lithium near the anode throughout the bath so that it can react effectively with the carbon dioxide released by the carbonate ion reaction at the cathode. It has been noted experimentally that the lithium oxide does tend to concentrate near the cathode and to form a pasty mass because its melting point is higher than the operating temperature. This mass tends to interfere with separation of the carbon as a dense phase at the cathode, and to entrain some of the salt electrolyte in the carbon precipitate. The entrained salt must be separated or discarded as expendable when the carbon is cleaned from the cell.

It should be emphasized here that the series of reactions presented to describe the mechanism of the operating cycle that has been demonstrated experimentally are only hypothetical. If operation were to depend on the postulated interfacial reaction of carbon dioxide gas in contact with liquid lithium metal, when oxygen gas is also present, success would appear impossible. The preferential reaction of atmospheric oxygen would be much more probable, to the exclusion of the carbon dioxide reaction, when both are present in a gas mixture. The experimental demonstration of the deposition of solid carbon in this system implies that the activity of oxygen is low compared to lithium and CO$_2$. This may result because both lithium metal and gaseous CO$_2$ are actually soluble in the melt, whereas oxygen is not.

The potential toxicity of the system has not been carefully evaluated. Presumably carbon monoxide might be a by-product under some conditions. The number of process steps has reached the irreducible minimum of one. Thus, the process receives a maximum rating on this count, which is counterbalanced by the complexity of the system chemically, as indicated briefly earlier.

COMPARISON OF SYSTEMS FOR CARBON DIOXIDE MANAGEMENT

At various points in the development of any integrated system for carbon dioxide management it is profitable to pause and to attempt a comparison of the various candidate systems on a common basis. Since newer concepts are invariably in an earlier stage of development, quantitative comparisons must always include a subjective appraisal of future progress that may eventually bring all systems into optimum and useful operation.

In a thorough and detailed study of the various factors that contribute to weight and power requirements for reducing carbon dioxide by either hydrogenation or electrolysis, the Hamilton Standard Division of United Aircraft Corporation has estimated the relative total weights of the various systems, including fixed weight, expendables, and a power-equivalent weight of 500 lb./kw., for six men on a 420-day mission. Their estimate indicates that the molten carbonate system would total about 900 lb., and the Bosch, with various alternate methods of CO_2 absorption and concentration, might represent somewhat less than 1,200 lb. Total weight of the solid electrolyte system would be of the order of 1,000 lb. These estimates are released with reservations as to their absolute accuracy, and represent only expert judgment rather than an established and proven capability for each process. The only firm conclusion from these estimates is that both the solid and the molten-salt electrolysis systems justify further examination because of their promising potential.

CONVERSION OF CARBON DIOXIDE IN BIOREGENERATIVE SYSTEMS

Bioregenerative systems are those in which the carbon dioxide is assimilated and converted by living organisms as part of their growth processes. This discussion presents only an elementary review of the significant characteristics of one such system. Most of the advantages and problems outlined here are characteristic of many bioregenerative systems, but the importance of each may vary over a wide range from system to system, depending on the specific growth requirements of the organism being used. This information may serve as a partial orientation for those whose main interest and experience is in the chemical conversion methods for CO_2 reviewed in the earlier sections. It also is intended to call attention specifically to one problem: Oxygen recovery from CO_2 by biosystems is subject to the constraints imposed in achieving simultaneous balances of both a carbon and a nitrogen cycle in order to maintain efficient growth. These constraints are well known to those studying biosystems for possible application to atmosphere management in space cabins. However, they have not often been discussed explicitly in this context. In addition, they have been sometimes ignored or de-emphasized in describing the potential use of combination oxygen-food cycles, based on utilization of human wastes (including CO_2) as nutrients for the organisms.

Figure 1 shows a concept of an oxygen-food cycle based on the continuous culture of *Hydrogenomonas* bacteria, which is being

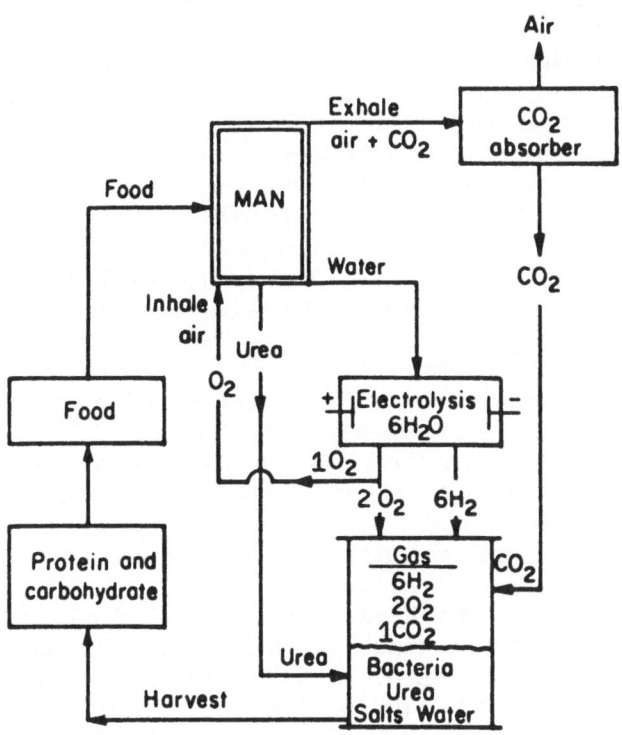

Figure 1.   Oxygen-food cycle for life support
with *Hydrogenomonas* bacteria culture.

studied in the author's laboratory (Foster and Litchfield, 1964) and
by others under NASA sponsorship. The hydrogenomonads assimilate
the three gases—hydrogen, oxygen, and CO_2—in the proportions
shown on the diagram. They can utilize urea as the sole nitrogen
source in the liquid medium, and CO_2 as the sole carbon source, except for carbon in the urea molecule. These are chemosynthetic
organisms that grow without illumination, but they are not light
sensitive. Hydrogen is utilized as a reductant for CO_2 in the growth
process that produces cell substance. Hydrogen also supplies process energy ultimately by reaction with the assimilated oxygen to
produce water.

   The bacterial harvest must be separated from the liquid
medium for possible use as food. Centrifuging is the conventional
procedure which is gravity insensitive. Culture in a liquid medium
is gravity sensitive, but various alternatives to rotating the whole
culture vessel could be considered when engineering development is
justified. The optimum culture temperature is about 90°F., and the

optimum pressure has not been determined. Pressure may be above cabin ambient to help prevent contamination of the culture by entrance of foreign organisms, but gas leaks must be avoided because of the flammability of hydrogen.

Suitability of the cell harvest for food has not been determined. Preliminary analyses indicate a protein content of over 50%, perhaps as high as 75%. All the amino acids necessary for human nutrition have been found. The possibility of formation of traces of toxicants in a steady-state culture seems unlikely, but definitive results are not yet available. Three steps —CO_2 collection, CO_2 conversion, and water electrolysis—are necessary in the integrated system for the oxygen cycle. These are the same as the three steps of the Bosch system, with the continuous culture substituted for the Bosch reactor. The difference is that potentially useful cell material is recovered instead of carbon.

The composition of the harvested cell material, as well as the growth rate, varies with the growth environment. Under the conditions for optimum growth so far identified in the studies of continuous cultures, the harvested cells contained about 48% carbon and about 12.5% nitrogen.

The CO_2 output of man varies considerably, depending on a wide variety of physiological conditions. It is assumed for this evaluation that the average CO_2 output is 1,000 gm. per man-day, representing a level of about 150% of the basal metabolic rate. When all of this CO_2 is assimilated by the growing culture, the harvested cells would contain $1,000 \cdot 12/44 = 273$ gm. of carbon in the cell substance. This is equivalent to $273/0.48 = 570$ gm. total dry weight of cell substance harvested per man-day.

To attain the nitrogen balance required for optimum growth, urea nitrogen (or nutrient nitrogen from another suitable source) would be assimilated at $570 \cdot 0.125 = 71$ gm. per man-day. This is equivalent to $71 \cdot 60/28 = 152$ gm. urea per man-day.

The urea output of man also varies considerably, and depends primarily on the amount of protein in the diet. The normal range is about 25 to 45 gm. of urea for a balanced diet containing 100 to 125 gm. of protein per man-day. Thus the nutrient nitrogen available from human waste is about 16 to 20 gm. nitrogen per man-day. This waste nitrogen is not adequate to supply the 71 gm. of nutrient nitrogen required to maintain optimum growth in the quantities required for complete CO_2 conversion, even when the relatively minor quantity of fecal nitrogen is also considered as a supplement to the urea nitrogen.

Perhaps some alternate nitrogen input could be made available for the growing culture. If so, the cell harvest from balanced CO_2 conversion by continuous culture would presumably contain much more protein (300 to 400 gm. protein per man-day) than would normally be useful as a dietary component. It is beyond the scope of this discussion to consider in detail the many combinations of food and oxygen cycles that might be devised for a closed system in which the oxygen consumed from the cabin atmosphere is regenerated partially or wholly with the use of bacterial cultures. Schlegel (1964) has suggested a search for oxyhydrogen gas bacteria that would store starch and glycogen in the cells, thus assuring a normal carbohydrate-rich food source suitable for a human nutritional balance. An obvious alternative would be conversion of part of the human CO$_2$ output by chemical methods, supplemented by CO$_2$ plus nitrogen-waste conversion in a balanced hydrogenomonad culture.

REFERENCES

Chandler, H. W., and W. Oser (1962). Study of electrolytic reduction of carbon dioxide. Tech. Doc. Rept. No. MRL-TDR-62-16 (March).

Foster, J. F., and J. S. McNulty (1961). Study of a carbon dioxide reduction system. ASD Tech. Rept. 61-388 (August).

_____, and J. H. Litchfield (1964). A continuous culture apparatus for the microbial utilization of hydrogen produced by electrolysis of water in closed-cycle space systems. Biotech. Bioeng., 6:441-456.

Remus, G. A., R. B. Neveril, and J. D. Zeff (1963). Carbon reduction system. Tech. Doc. Rept. No. AMRL-TDR-63-7 (January).

_____, _et al._ (1965). Catalytic reduction of carbon dioxide to methane and water. Tech. Rept. No. AFFDL-TR-65-12 (April).

Rousseau (1963). Atmospheric control systems for space vehicles. Tech. Doc. Rept. No. ASD-TDR-62-527, Pt. I (March).

_____(1964). Atmospheric control systems for space vehicles. Tech. Doc. Rept. No. ASD-TDR-62-527, Pt. II, pp. 101-118 (February).

Schlegel, H. G. (1964). The use of H$_2$-oxidizing bacteria to regenerate respiratory air. Raumfahrtforschung, 8(2):65-67 (NASA trans. TT F-8899).

Thompson, E. B. (1964). Investigation of catalytic reaction for CO$_2$ reduction. P. I: Evaluation of a nickel kieselguhr catalysis. Tech. Doc. Rept. No. FDL-TDR-6422, Pt. I (October).

_____. (1965). Pt. II: Evaluation of base metal oxide catalysis. Tech. Doc. Rept. No. FDL-TDR-6422, Pt. II (April).

6.

GASEOUS DIFFUSION CELLS

COLEMAN J. MAJOR · RICHARD W. TOCK

The University of Akron, Akron
State University of Iowa, Iowa City

INTRODUCTION

The removal of carbon dioxide and other noxious vapors from the life-support system of a spacecraft is a formidable problem in manned space exploration. This is particularly true for flights of long duration, and in those instances when a significant number of astronauts are involved in the flight. Of the several different systems proposed, the gaseous diffusion, or permeability, process incorporates three attractive features.

1. It utilizes a barrier having a selective permeability with respect to carbon dioxide. Since it operates entirely in the gaseous phase, the driving force is a partial pressure gradient. The system therefore, is independent of gravity, and is unaffected by the conditions of weightlessness.
2. The gaseous diffusion system is a continuous operation. As such, the selective barriers need not be recharged or regenerated in order to continue to effect a separation.
3. The membranes or selective barriers of such systems are chemically inert and stable over a relatively wide temperature range.

The gaseous diffusion process has, however, two principal disadvantages.

1. It requires relatively high pressure drops for most selective barriers. This means that the power requirements for interstage compression could be prohibitive for the power systems of small spacecraft.
2. A diffusion system cannot provide a complete separation in a single pass. A cascade system would therefore be necessary

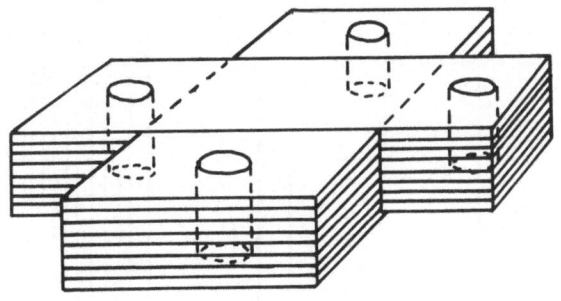

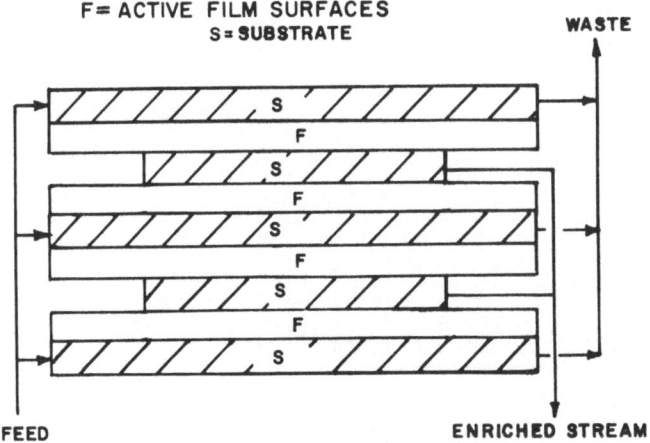

Figure 1. Graphic representation of cell assembly; F, active film surfaces; S, substrate.

in order to provide adequate separation of the individual components. The volume requirements for a cascade system could become excessive for spacecraft use.

One of the classical methods of obtaining a large surface area
per unit volume is that exhibited by a tube bundle in a heat exchanger. For the proposed new design, alternating thin layers of
a selective plastic membrane and a porous substance will replace
the tubes in the heat exchanger model. Figures 1 and 2 show the de-
tails of the cell construction. The unique feature of this design is
the manner in which interconnection between alternate compartments is accomplished. This technique permits hundreds or thousands of individual cell compartments to be connected in parallel

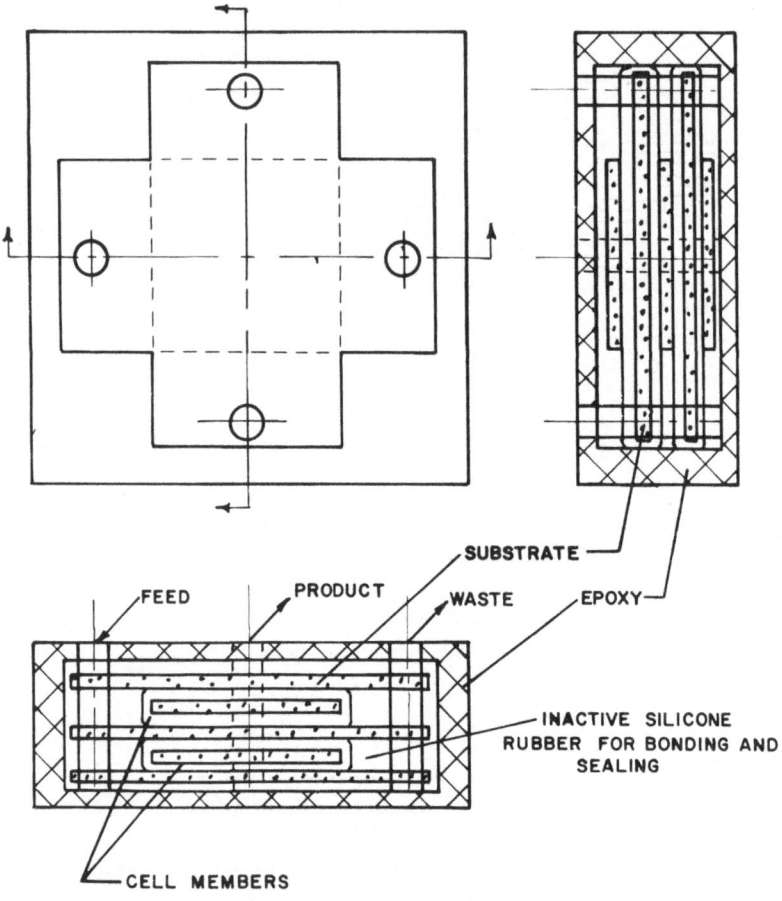

Figure 2. Sectional view of potted cell.

without complicated assembly operations. The design also lends itself to construction of rigid hermetically sealed units similar to the potted components used in the electronics industry.

Before a successful cell could be built, however, a number of variables had to be studied, and several manufacturing techniques had to be developed. To accomplish this, the plan of study summarized in the following paragraphs was proposed and followed.

Every effort was made to ascertain the materials best suited for cell construction. Permeability tests were made on a number of selective membranes, as well as on samples of the porous medium that was to be used between the successive layers of the selective membranes. The latter permeability tests were conducted with edgewise flow instead of the usual right-angle flow. This helped to determine whether the pressure drop through that particular porous medium would be excessive.

Beginning with the best construction materials, techniques were developed for coating the porous medium with a thin film of the plastic membrane. Tests were then devised to determine the minimum practical thickness for the laminated film. Permeability tests of the film laminated to the porous medium were also conducted, the principal point of which was to make sure that a continuous film had been formed and that the polymer had not penetrated the porous medium sufficiently to clog its pores.

Experiments were conducted to determine the best procedures for assembling a cell unit. These included tests on the drilling operation for cutting the holes to interconnect the alternate layers, tests on procedures for bonding the individual layers to one another, and experiments with materials to be used for potting the cells. When a method for constructing the cell had been perfected, a hand-assembled unit was built and tested.

THEORY

The proposed permeability cell is a unit for separating gaseous mixtures by permeation through a semipermeable membrane. Since a separation is desired, the membrane must demonstrate a different permeability for each of the gaseous components being separated. Moreover, for a separation by permeation it is essential that every membrane used be completely nonporous. If the membrane contains pinholes or if other microscopic defects that destroy the continuity of the film are present, the selectivity of the film will be lost. Consequently, any resulting separation that is obtained will not be by permeation, but rather by diffusion or some other less selective process.

For a single-stage unit as shown in Figure 3, a permeability separation requires four basic steps.

1. The gaseous mixture must be brought into contact with one side of the semipermeable membrane. The contact between gases and membrane results in the absorption and dissolution of the gases, thereby causing a partial pressure gradient to develop across the membrane.

2. Under the influence of the partial pressure gradient, the gases will begin to permeate the membrane. The direction of permeation is through the membrane from the high pressure side to the low pressure side. The rate of permeation for each component of the mixture will be governed by the selective properties of the film and by the magnitude of the partial pressure gradient.

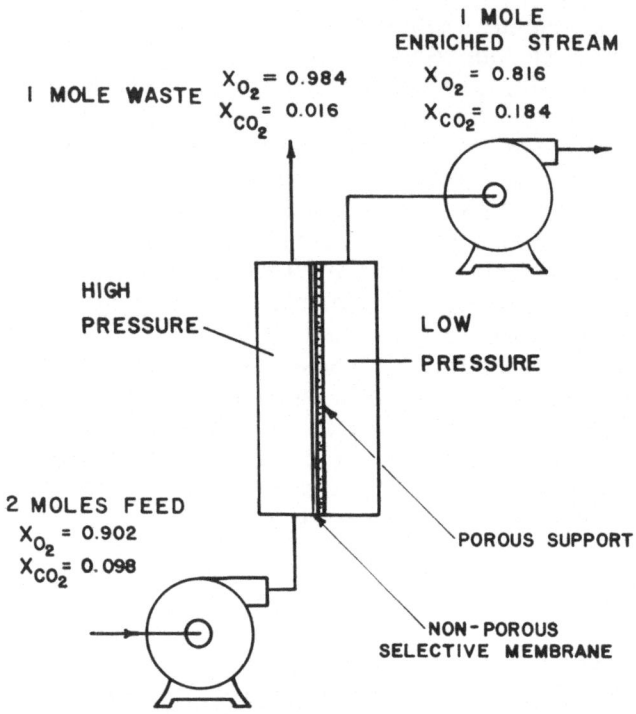

Figure 3. Schematic of a single-stage unit for separation by selective permeation.

3. The permeated gas, now enriched in at least one of the components, is evaporated from the low pressure side of the membrane and removed as product from the system. The removal of the product stream is necessary to sustain the partial pressure gradient across the membrane.

4. The unpermeated and depleted gas is similarly removed as waste from the high pressure side of the system.

The gas permeability coefficient for a film is defined as the volume of a pure gas flowing normal to two parallel surfaces a unit distance (thickness) apart, under steady conditions, through unit area under unit pressure differential, at a stated test temperature. As such, the coefficient is a basic property of the material and independent of specimen geometry. It is related to the diffusion rate and to the solubility of a gas by the equation

$$\overline{P} = DS$$

where P = gas permeability coefficient, D = diffusion rate, and S = solubility. An accepted unit of $\overline{P}$ is cc. (at standard conditions)/ sec., sq. cm., cm. of mercury (pressure)/cm. of thickness at the stated temperature of the test. A convenient unit is the *barrer*:

$$\overline{P} \text{ (barrer)} = \frac{10^{-10} \text{ cc. } (STP) \text{ cm.}}{\text{sec.-sq. cm.-cm. Hg}}$$

In order to determine the relative rate of separation of a binary mixture, one must first find the ratio of the gas permeability coefficients for the temperature at which the separation is to take place. If this ratio is greater or less than unity, a separation of the mixture is possible, and the rate of separation will be proportional to the relative magnitude of the ratio. If the conditions are such that the ratio obtained is equal to unity, then the mixture will not separate, but will behave as an azeotropic mixture.

The theory for calculating the enrichment of a binary mixture has been well developed by Weller and Steiner (1950) for the single-stage permeability operation. The Weller Steiner Case I for perfect mixing of the gases on both sides of the film was chosen to describe the proposed permeability cell. Using this model it is assumed that the effective composition on the high pressure side of the selective membrane will be equal to the composition of the gas leaving that region. Correspondingly, the composition on the low pressure side of the membrane will be the same as the gas stream leaving that region.

Based on the foregoing assumption and on Fick's law of diffusion, Kammermeyer and Brubaker (1952) derived equations for binary, ternary, and quaternary systems. The following equation represents the equilibrium dynamics for the binary system:

$$\frac{(X_p)}{(1 - X_p)} = R \frac{(\mathscr{P} X_0 - P X_p)}{[\mathscr{P} (1 - X_0) - P(1 - X_p)]} \tag{1}$$

where (X_p) is mole fraction of component A in the product stream; $(1 - X_p)$, mole fraction of component B in the product stream; (X_0), mole fraction of component A in the waste stream; $(1 - X_0)$ mole fraction of component B in the waste stream; R, the ratio of the pure gas permeability coefficients, for gases A and B (separation factor = $\overline{P}_A / \overline{P}_B$; $\mathscr{P}$, the total pressure on the high pressure side of the membrane in cm. Hg; P, the total pressure on the low pressure side of the membrane in cm. Hg.

Equation 1 is a quadratic equation in terms of (X_p), and a solution for (X_p) can be written in the form

$$(X_p) = \frac{-b + \sqrt{b^2 - 4ac}}{2a} \tag{2}$$

The values of the quantities a, b, and c were obtained using over-all material and component balances. Based on (ϕ) the cut, and (X_F) the mole fraction of (A) in the feed stream, the following values were obtained:

$$a = (R - 1)(P + \phi \Delta P)$$

$$b = -(R - 1)(\mathscr{P} X_F + \phi \Delta P) - \Delta P - RP$$

$$c = R \mathscr{P} X_F$$

where $\phi = \text{cut} = \dfrac{\text{Product rate (cc./ min.)}}{\text{Feed rate (cc./min.)}} = \left\{ 1 - \dfrac{\text{Waste rate}}{\text{Feed rate}} \right\};$

$\Delta P = (\mathscr{P} - P)$ cm. Hg; X_F = mole fraction of component A in the feed stream. Typical separation curves derived from this theoretical equation are plotted in Figure 10

It is interesting that, for a permeability separation of a binary mixture, the degree of separation with respect to product purity will approach a maximum value as the amount of product removed approaches zero. Theoretically this implies that the composition of component A in the waste stream (X_0) approaches the composition of A in the feed stream (X_F) as the composition of A in the product (X_p) approaches a maximum. Moreover, the pressure on the product side of the membrane recedes to zero. Under these conditions, Eq. 1 reduces to

$$X_0 \to X_F, \qquad P \to 0, \qquad \text{and} \qquad X_p \to X_p \text{ (max)}$$

$$\frac{X_p \text{ (max)}}{(1 - X_p)(\text{max})} = \frac{RX_F}{(1 - X_F)} \tag{3}$$

Although these conditions produce the highest degree of product purity for the single-stage operation, they will not produce the greatest amount of product take-off and over-all separation. Therefore a system is seldom operated at these conditions.

Furthermore, that the purity of the product does approach a finite maximum, or upper limiting value, is evidence that a permeability separation cannot produce a complete separation in a single pass. This can be accomplished only with a membrane having an infinite separation factor $(R = \infty)$. Unfortunately at present there is no known membrane having a separation factor greater than 10 for a mixture of carbon dioxide and oxygen. A great many membranes have separation factors of less than 10 for this mixture, however, and these offer the possibility of very nearly complete separations when employed in cascade arrangements.

In those instances when X_F becomes quite small, the quantity $(1 - X_F)$ will be approximately equal to unity. Under these conditions Eq. 3 may be further reduced to

$$X_p \text{ (max)} = \frac{RX_F}{1 + RX_F} \tag{4}$$

APPARATUS AND PROCEDURE

This investigation consisted of three distinct phases. The first phase involved an investigation of the materials and techniques needed for the construction of a permeability cell. The second phase revolved around the construction of a prototype of the cell. The final phase dealt with tests to determine the operating characteristics of this prototype. Each phase of the investigation required a number of different apparatus arrangements.

Film preparation and testing

Silicone rubber samples were mixed according to the instructions recommended by their respective manufacturers, Dow Corning and General Electric. A Gardner precision drawing knife that could be adjusted to various thicknesses was used to form films on clean plate glass. For this operation it was essential that the glass be completely free of scratches and dust particles to insure continuity of the film and to facilitate its removal after polymerization.

The selective properties of the silicone films were tested with a compact gas-permeability apparatus designed and produced by the C. J. Major Co. (1963). Figure 4 is a schematic representation of the apparatus. The unit has a durable construction, is completely portable, and does not require a vacuum source, utilities, or special tools for assembling and disassembling. The permeability coefficient of the film being tested is determined by timing the displacement of a water slug along a calibrated capillary tube. Results obtained with this apparatus correlated well with those of larger and more elaborate arrangements.

A significant amount of permeability data covering a wide range of different polymers and several gases had been accumulated by Major, Kammermeyer, Roberts, and McIntosh (Major and Kammermeyer, 1962; Major, 1963; McIntosh, 1963; Roberts, 1962)

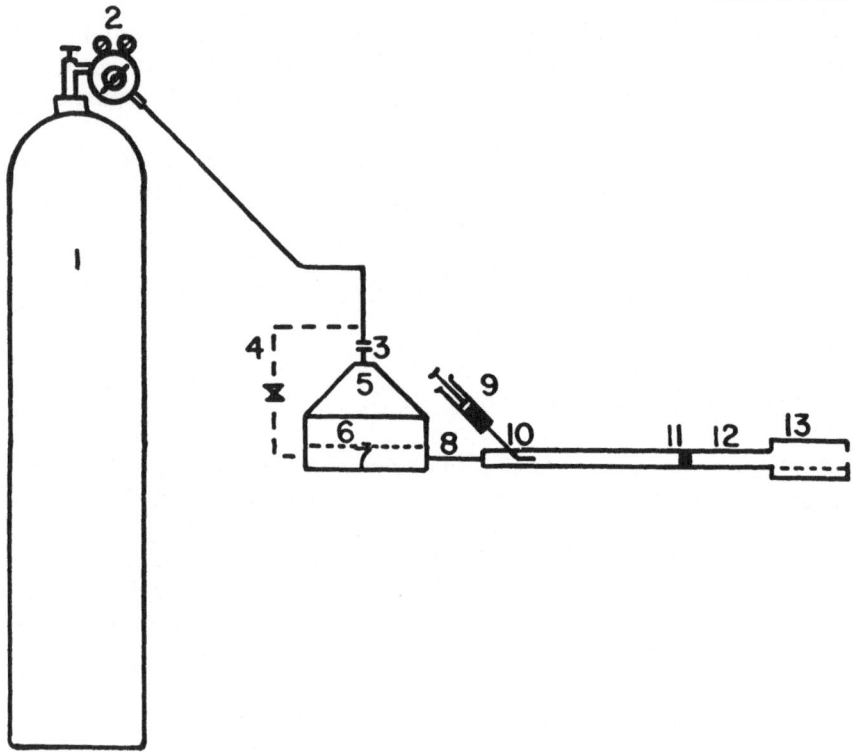

Figure 4. Schematic representation of compact gas-permeability apparatus designed and produced by the C. J. Major Co.

prior to this investigation. McIntosh also extended the permeability data for several silicone rubbers over a temperature range of 40°C. to 10°C. (McIntosh, 1963). All these sources were useful in the final analysis and selection of the cell membrane.

Substrate testing

A preliminary study of the structure of the proposed permeability cell revealed the need for a substrate that would separate and support the semipermeable films. In addition, these substrates were to provide channels through which the gases could travel in going to and from the film surfaces. Since it was desirable that these substrates offer a minimum of resistance to the gas flow, the porosity of a material became the primary criterion for judging its excellence as a substrate. An indication of the porosity of potential substrate materials was obtained with the apparatus depicted in Figures 5 and 6.

This apparatus was designed to measure the volumetric flow

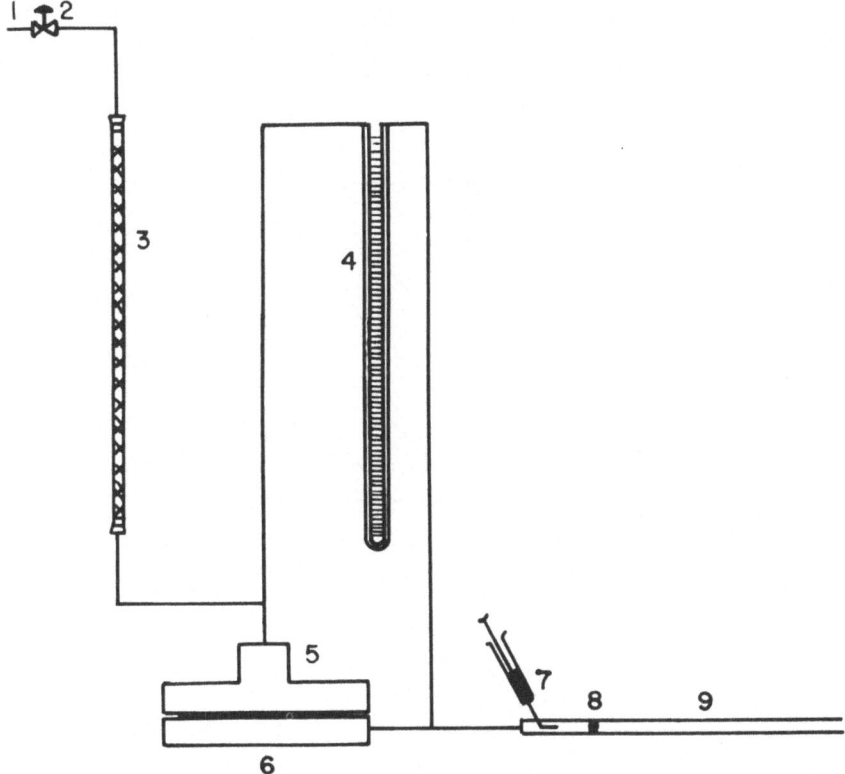

Figure 5. Schematic of substrate porosity apparatus.

rate of air edgewise through the plane of the sample. As shown by the schematic diagram, this was accomplished by measuring the rate of slug travel through a capillary tube in a manner similar to that for the compact permeability apparatus.

Figure 6 depicts an exploded view of the unit for holding a sample of the substrate materials. The body of the assembly is constructed of machined Lucite, and the separate halves are held together by a bolt through each of the four corners. Air enters and leaves the cell through $\frac{1}{4}$-in. flare couplings. The substrate material is held in place by pressure exerted by compression of a spring. Air is forced to flow radially from the periphery of the sample to its center by the bottom portion of the cell and a machined aluminum disk resting on top of the sample. As an added precaution, silicone films were placed above and below the substrate sample. These films acted as gaskets and minimized gas flow by channeling across the sample's surface. Thus a better picture of the material's actual porosity was obtained.

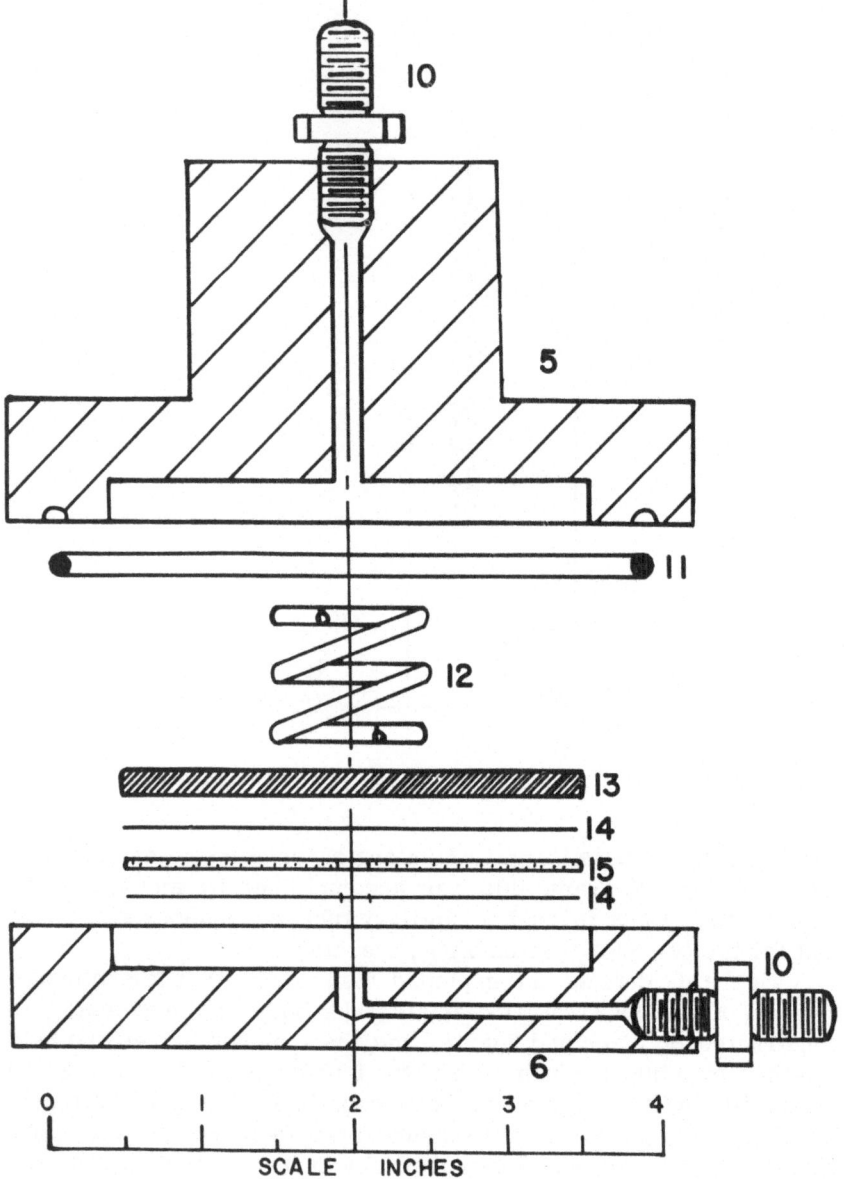

Figure 6. Substrate sample holder of porosity apparatus depicted in Figure 5.

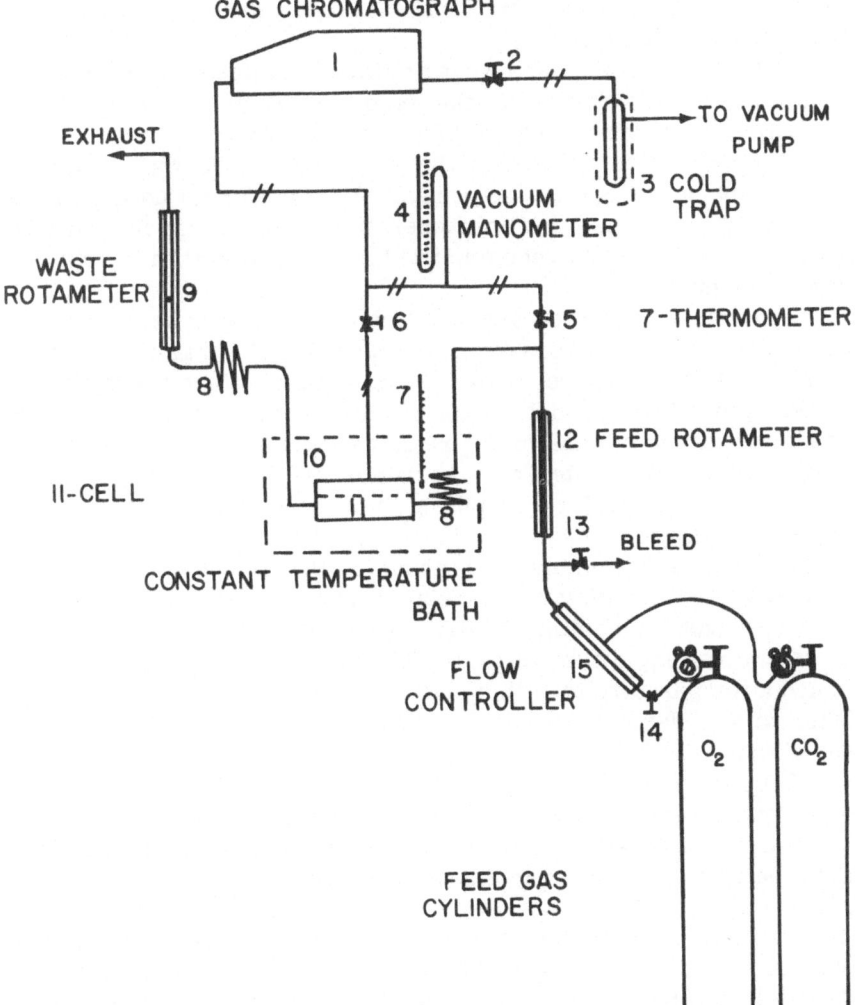

Figure 7. Schematic of apparatus used to test the separation produced by a cell.

Cell testing

Prototypes of the permeability cell were tested to determine their ability to effect a separation and to measure their handling capacity. Figure 7 is a schematic representation of the apparatus used for testing the separation produced by a cell.

A special gas proportioning valve was used to produce a continuous flow of a constant gas mixture. This instrument has excellent application over the small carbon dioxide concentration ranges used in the test.

A Fisher Porter and a Roger Gilmont rotameter were used to measure the flow rates of the feed and waste streams. Their capacities ranged from 0 to 260 std. cc. air/min.

The driving force across the cell membranes was produced by creating a vacuum on the product side. A Welch Duo-Seal vacuum pump was used for this operation. A cold trap for volatile collection was used as recommended in connection with the operation of the pump.

Analysis of the feed and product streams was accomplished using a special vacuum setup on a Beckman GC-2 gas chromatograph. A 4-ft. silica-gel column was used to effect a partition of the sample within the instrument. A closed-end Fisher manometer was used to record the pressure on the product side of the membranes. This manometer also served to record sample sizes, and thereby provided uniform sampling in connection with the chromatographic analyzer.

A Sargent thermometer, model S. W., temperature-control bath was used to maintain a constant temperature during each run. This unit has its cooling, heating, and agitation systems all integrated into a single control panel. When properly adjusted it maintained a temperature to within $\pm 0.01°C$. Temperature readings were taken with a Fisher thermometer having a range of $-36°C$. to $54°C$., and divisions of $0.2°C$.

Copper tubing ($\frac{1}{4}$-in. O.D.) and sweat fittings were used for all lines subjected to vacuum pressures. Copper conditioning coils were also employed to help bring the gas streams to equilibrium upon entering and leaving the bath portion of the system. Tygon tubing (3/16-in. O.D.) was used to make all the remaining connections within the system. Vacuum grease was used to help seal these connections. To check the capacity of the substrates within the cell, a simple setup using a mercury manometer and a wet test meter were used.

DISCUSSION OF RESULTS

Eight different formulations of silicone elastomers were tested. The results of the tests are presented in Table 1. From these data it is apparent that each type of elastomer exhibited approximately the same 5:1 ratio between the carbon dioxide and the oxygen

TABLE 1

PERMEABILITY COEFFICIENTS—SILICONE RUBBER

$$\bar{P} = \frac{cc.(STP)(cm. \times 10^9)}{sec.-sq.cm.-cm.Hg}$$

Silicone rubber membrane	Temp. (°C.)	Permeability constant			Separation factor	
		CO_2	O_2	N_2	RCO_2-O_2	RCO_2-N_2
RTV-501	23.0	285	53.8	25.8	5.30	11.05
	32.5	278	59.3	29.2	4.69	9.52
	43.0	280	65.7	33.4	4.26	8.38
	6.0	270	42.5	18.8	6.35	14.36
RTV-502	23.0	286	55.3	26.3	5.17	10.87
	33.0	280	60.4	29.9	4.64	9.36
	43.0	282	66.3	34.1	4.25	8.27
	10.5	269	45.9	20.7	5.86	13.00
RTV-40	24.0	205	42.6	21.4	4.81	9.58
	33.5	203	45.8	23.3	4.43	8.71
	43.0	197	46.8	24.1	4.21	8.17
RTV-11	29.0	240	50.6	24.6	4.74	9.76
	33.0	238	51.4	25.5	4.63	9.33
	43.5	235	57.1	29.0	4.12	8.10
RTV-601	33.0	286	75.6	45.0	3.78	6.36
	43.0	282.	77.7	47.5	3.63	5.94
RTV-20	28.5	191	39.9	18.8	4.79	10.16
	33.0	190	41.0	19.8	4.63	9.60
	43.0	189	45.6	25.2	4.14	7.50
Eccosil 4712	20.5	137	28.0	13.0	4.89	10.54
	32.0	138	31.2	15.2	4.42	9.08
	43.5	139	34.3	17.4	4.05	7.98
Sylgard 182	20.5	204	40.0	18.1	5.10	11.27
	33.5	206	46.1	21.7	4.47	9.49
	43.5	205	51.2	24.9	4.00	8.23

permeability coefficients at room temperature. The carbon dioxide to nitrogen ratio is about 9:1 for the same temperature. Actual permeability values for carbon dioxide range between 137×10^{-9} and 286×10^{-9} whereas those for oxygen range between 28×10^{-9} and 78×10^{-9}.

Figures 8 and 9 show plots of the permeability coefficient versus $(1000/T°K)$ for the silicone rubbers RTV-501 and RTV-502 with respect to the pure gases, oxygen, nitrogen, and carbon

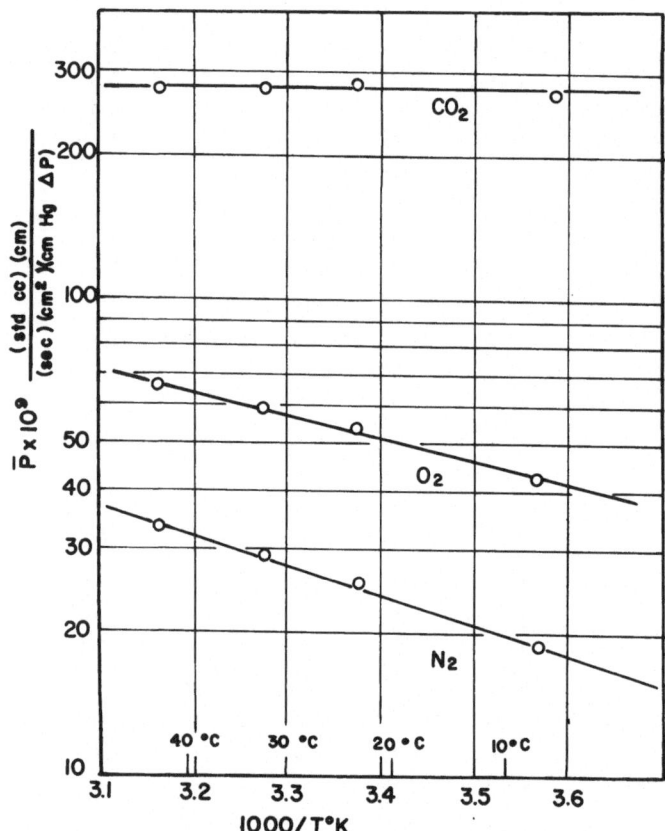

Figure 8. Plot of permeability coefficient for silicone rubber RTV-501.

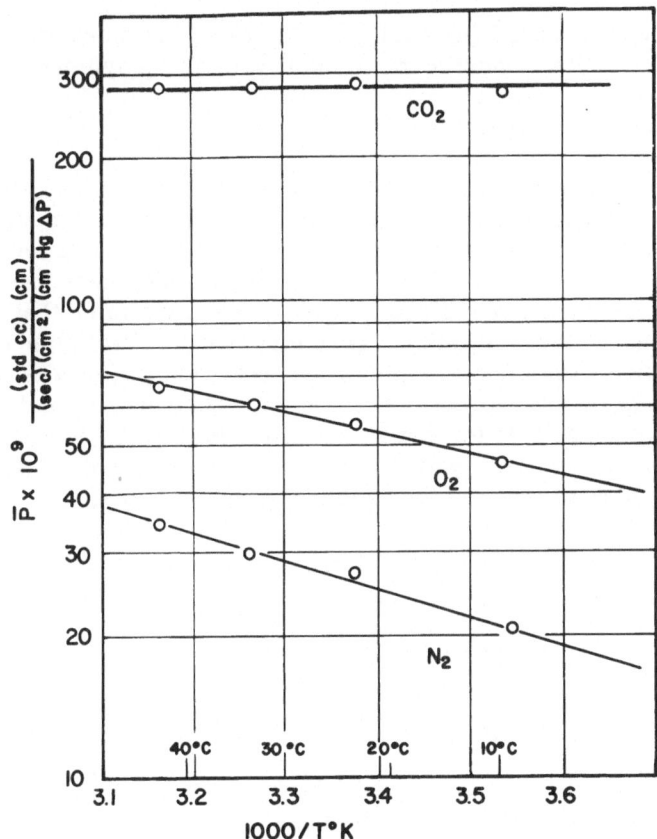

Figure 9. Plot of permeability coefficient for silicone rubber RTV-502.

dioxide. From these plots it is apparent that the permeability coefficients for these elastomers with respect to carbon dioxide is essentially independent of temperature. Permeabilities obtained for oxygen and nitrogen do, however, show a direct dependence on temperature. Since the oxygen permeability coefficient decreases with temperature while the carbon dioxide coefficient remains constant, the ratio $(\bar{P}_{CO_2}/\bar{P}_{O_2})$ must increase with decreasing temperature. These data would then indicate that the optimum operating conditions for a completed cell is somewhere in the lower temperature ranges, where the separation factor is the greatest.

With but two exceptions, the permeability data obtained are essentially the same for all the silicone rubber formulations that were tested. However, RTV-501 silicone rubber, manufactured by Dow Corning, was selected for use in cell construction. This selection was based primarily on its excellent workability in all phases and its superior film quality in the polymerized phase.

Substrates and porosity

Lined notebook paper, filter paper, Nibroc paper towel, kraft paper, Gelman Polypore, and typing paper were the materials selected for testing as potential substrates. The volumetric flow rate of air through the plane of each sample was measured. These rates were then normalized to give a porosity value that was independent of the specimen's geometry. Since a relative porosity was desired for each sample, lined notebook paper was taken as the standard. Its porosity reading was set equal to unity. Based on this standard, filter paper exhibited the highest porosity of the samples tested and typing paper the lowest. Table 2 shows the relative porosity values for each of the samples tested.

TABLE 2

SUBSTRATE POROSITY

Sample	Relative porosity	Texture
Lined paper	1.00	Glossy
Filter paper	11.80	Wrinkled, coarse
Nibroc paper towel	8.70	Coarse
Kraft paper	5.90	Rough
Gelman Polypore	4.40	Tight-woven cloth
Typing paper	0.46	Smooth, tight

Of the materials tested, Nibroc paper towel and kraft paper were selected, on the bases of their high relative porosity, availability, and workability, for further study and testing. Both substrates were laminated to RTV-501 silicone films, and the porosities of these laminated substrates were subsequently measured. The averaged results are tabulated in Table 3, together with their respective porosities for the unlaminated state. The significant decrease in the porosity of a sample after it had been laminated is clearly shown in this table. It was assumed that this porosity decrease was in part an indication that channeling had occurred across the surfaces of the unlaminated sample, and therefore, a true porosity value for flow through the sample had not been obtained. It was also assumed that a portion of the loss in porosity could have arisen from the penetration of the substrate by the

TABLE 3

LAMINATED SUBSTRATE POROSITY

	Relative porosity	
Sample	Laminated	Unlaminated
Lined paper		1.00
Nibroc paper towel	5.95	8.70
Kraft towel	2.36	5.90

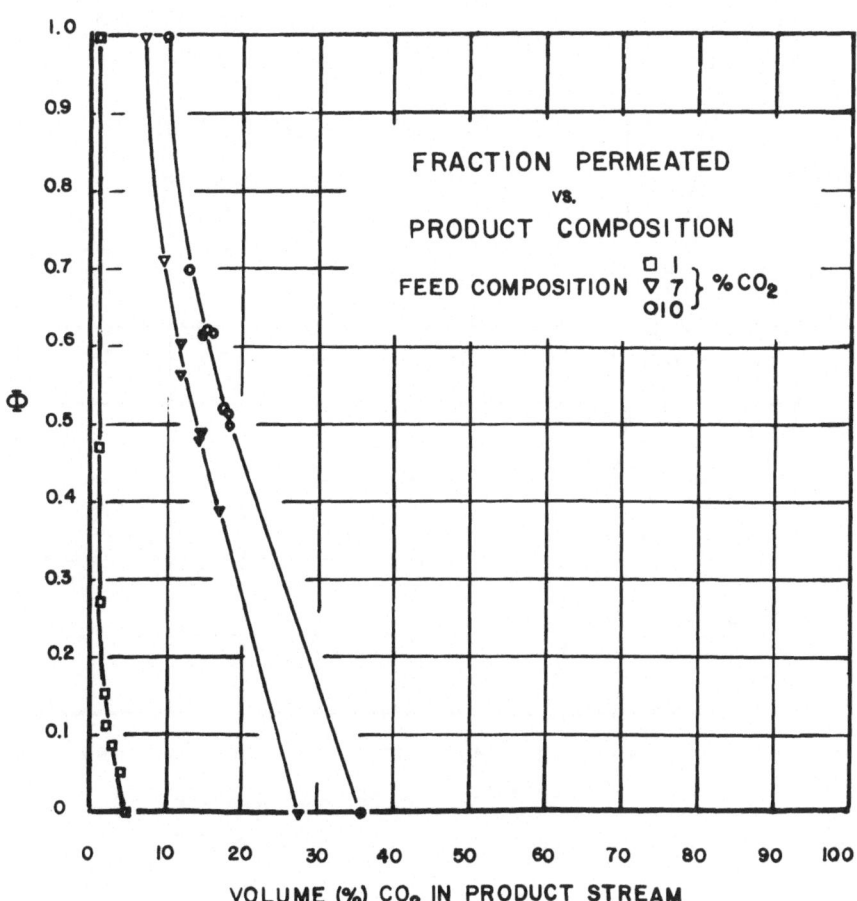

Figure 10. Separation curves for cell 3A at 40°C., O_2-CO_2.

tacky silicone rubber during the lamination process. Upon curing
it would then effectively plug those porous passages into which it
had penetrated. Since a substrate was desired that would least
restrict the flow of gases, the effects of any channeling that might
occur were considered somewhat desirable. Excessive penetration
of the substrate by the silicone was, on the other hand, considered
most undesirable and to be avoided.

Since the Nibroc paper had a relatively high porosity and
seemed to laminate well to the silicone films, it was selected for
use as a substrate in the construction of the cell.

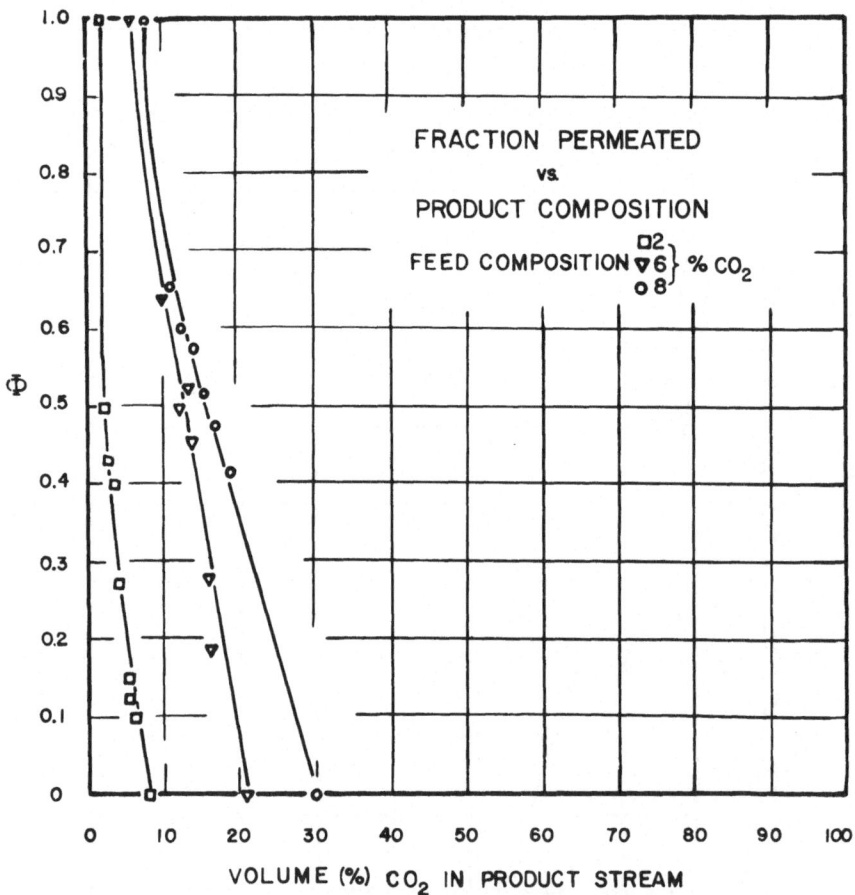

Figure 11. Separation curves for cell 3A at 25°C.

Cell assemblies

Nearly 30 prototypes of the proposed permeability cell were constructed and tested. Approximately one-fifth of these effected a satisfactory separation while the remaining four-fifths were not successful. For this investigation a cell producing a satisfactory separation was defined as one having a carbon dioxide to oxygen ratio greater than 4.8 for its separation factor. A large portion of the cells that were rejected on this basis had separation factors ranging from 4.5 down to 2.0.

The results of the tests performed on a satisfactory cell, 3A, are shown in Figures 10 to 12, where the separation curves for the cell at three different temperature isotherms, 40°C., 25°C.,

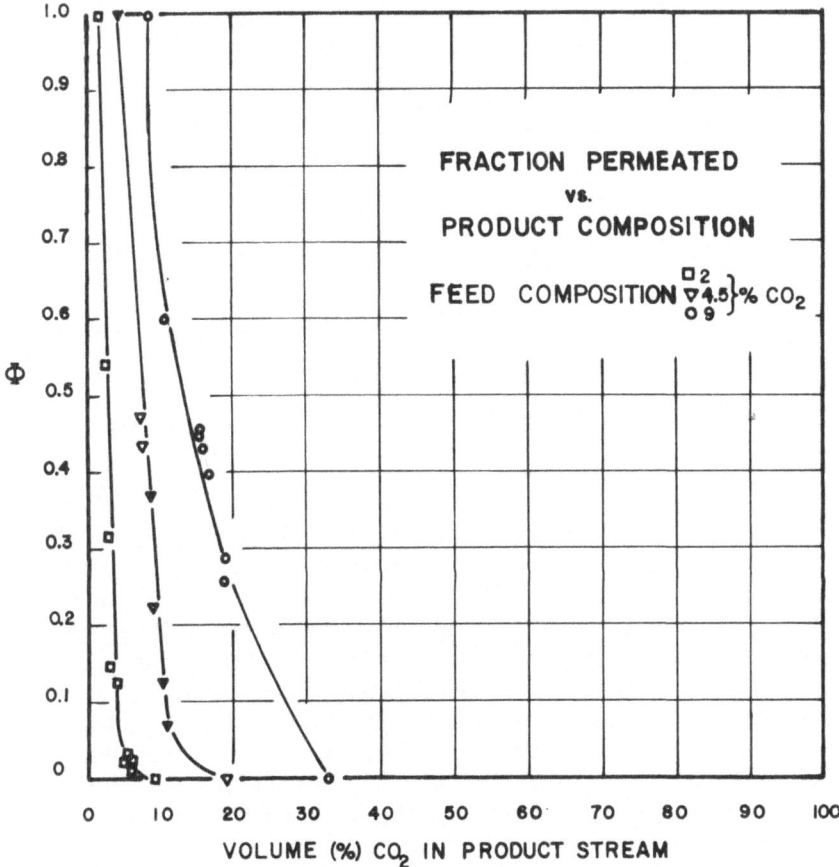

Figure 12. Separation curves for cell 3A at 0° C.

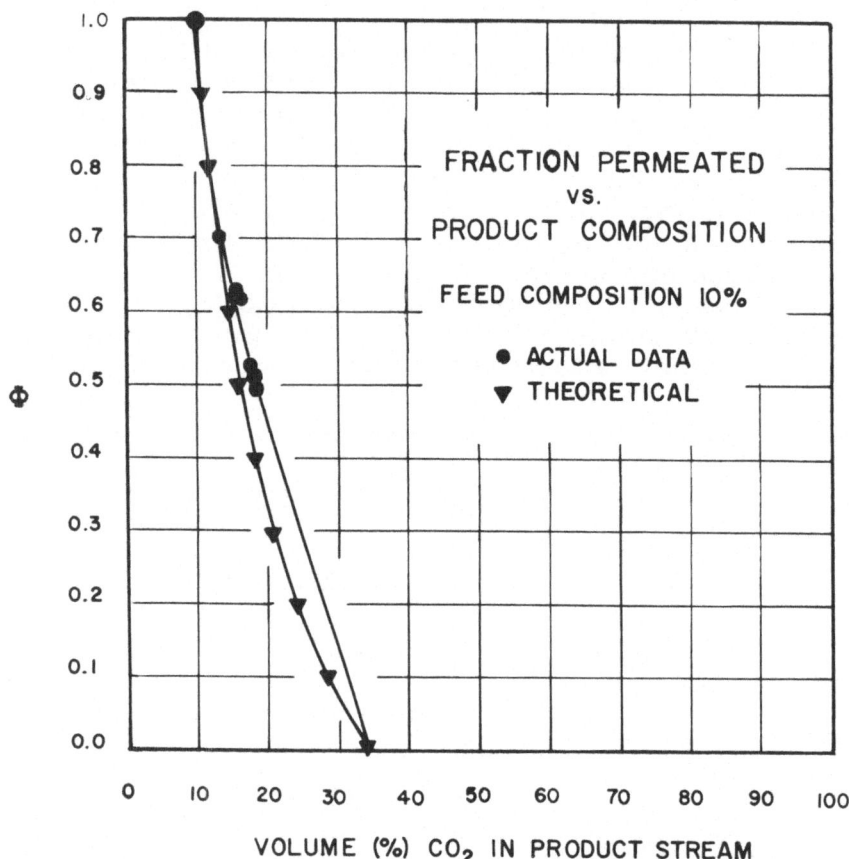

ϕ

Figure 13. Actual and theoretical separation curves: cell 3A; 40° C.; O_2-CO_2.

and 0°C., and for various feed compositions, ranging from 1% to 10% carbon dioxide in oxygen, are presented. The curves are plotted as cut (ϕ), the ratio of product rate to feed rate, versus the composition of the product stream. The end points of these curves represent the two possible extremes for a separation produced by this cell. When the fraction permeated (cut ϕ) is equal to 1.0, there is no separation and the product must of necessity be equal to the feed in both composition and rate of flow. As (ϕ) approaches zero the amount of product taken off also approaches zero. At this point the maximum product purity for the separation is achieved. The product composition for this point was predicted from the simplified form of the Weller Steiner Case I equation. It can also be approximated graphically by extrapolating the curve to (ϕ) equals zero and reading the value of the product composition.

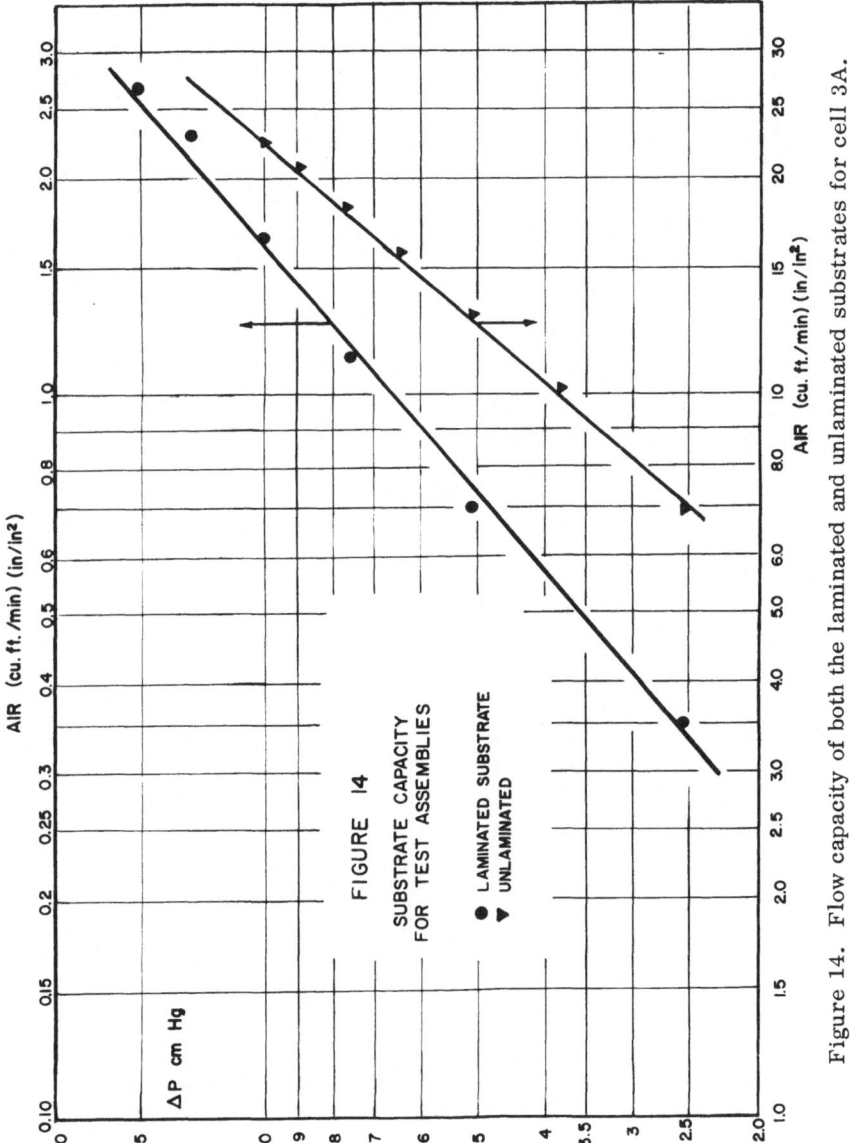

Figure 14. Flow capacity of both the laminated and unlaminated substrates for cell 3A.

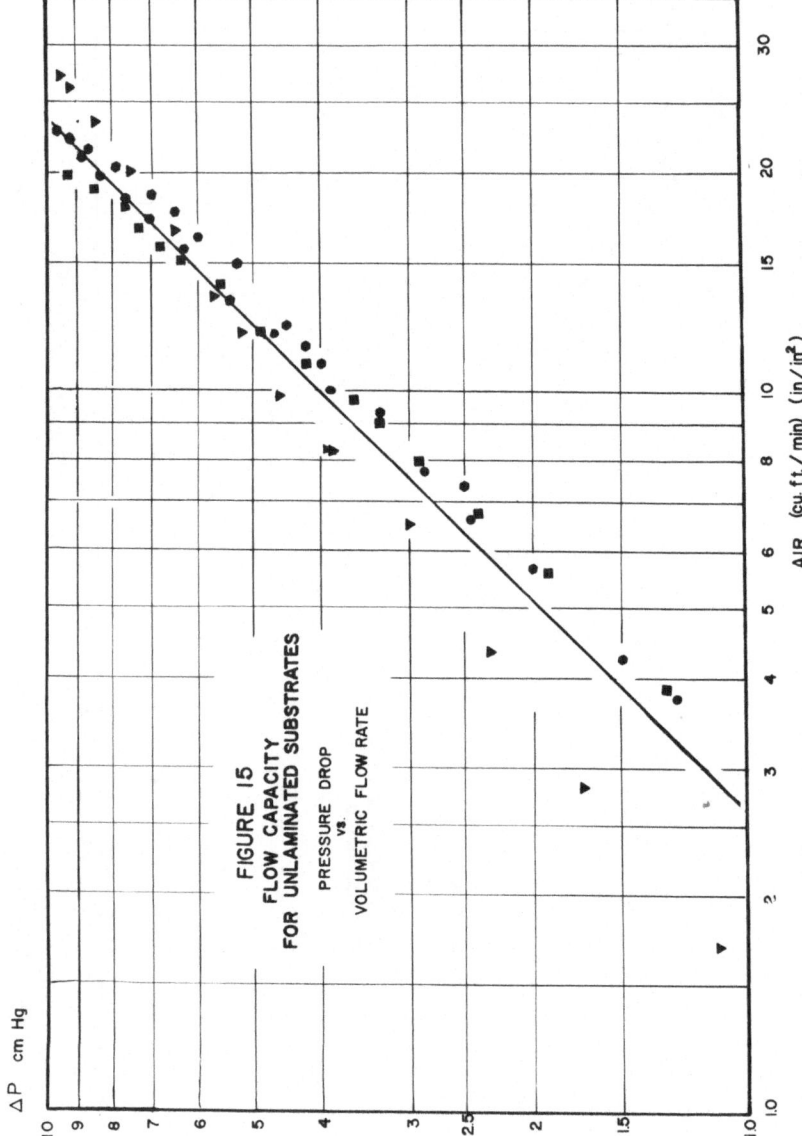

Figure 15. Flow capacity for unlaminated substrates for several cells.

The graph in Figure 13 depicts the separation curve derived from the theoretical considerations of the Weller Steiner Case I equation as opposed to that described by the actual data. It is evident from this plot that the Case I equation does not hold for the entire range of (ϕ), but begins to deviate at a value of (ϕ) between 0.70 and 0.80. This would seem to indicate that perhaps perfect mixing was not occurring on the high pressure side, and that perhaps the Weller Steiner Case II equation for laminar flow is more applicable for describing the system. The constant temperature runs at 0°C. (see Figure 12) with their steep linear slopes do, however, have a closer resemblance to the Case I predictions. It was theorized that this was caused by the decreased rate of total permeation at these low temperatures. This would subsequently permit the gases to mix more without being absorbed, which is of course one of the primary conditions that must be met to satisfy Case I.

The cell 3A used in the test had an over-all volume of 345 cc. $(2.54 \times 11.42 \times 11.42)$, which includes the potting material. This volume contained a combined film surface of 640 sq. cm. having an average thickness of 0.0206 cm. (8 mil). Figure 14 depicts the flow capacity of both the laminated and unlaminated substrates for this cell. Figure 15 shows the capacity of the unlaminated substrates for several cells. The maximum pressure drop exerted across the membranes was approximately 760 mm. Hg. On the high pressure side the feed was admitted and the wastes expelled at slightly above atmospheric pressure. The low pressure side was kept at 10 mm. Hg. vacuum throughout the test run.

Nomenclature

R = ratio of pure gas permeabilities $(\overline{P_A}/\overline{P_B} = R_{AB})$

a = constant for solution of Weller Steiner Case I equation

$$(R - 1)(P + \phi \Delta P)$$

b = constant for solution of Weller Steiner Case I equation

$$[(R - 1)(-\mathscr{P}X_F - \Delta P) - \phi \Delta P - RP]$$

c = constant for solution of Weller Steiner Case I equation

$$\mathscr{P}RX_F$$

ΔP absolute pressure drop across the cell membranes $(\mathscr{P} - P)$
= p.s.i.a.

$\mathscr{P}$ pressure on high pressure side of the membrane (p.s.i.a.)
P pressure on low pressure side of the membrane (p.s.i.a.)
$\overline{P}$ permeability constant (std. cc.-cm./cm.-sec.-cm. Hg)
ϕ ratio of product flow rate to feed flow rate or the cut
X_F mole fraction of carbon dioxide in feed stream
X_0 mole fraction of carbon dioxide in waste stream
X_p mole fraction of carbon dioxide in product stream

Acknowledgment

This work was performed under the sponsorship of the National Aeronautics and Space Administration under Contract No. NASr-73. The authors gratefully acknowledge the support given them by that agency.

REFERENCES

Brubaker, D. W., and K. Kammermeyer (1952). Ind. Eng. Chem., 44: 1465-74.
Major, C. J. (1963). Modern Pkg., 36 (5): 119-122 (June).
——, and K. Kammermeyer (1962). Modern Plastics, 39 (11): 135-146; 179-180 (July).
——(1963). Interim Tech. Rep. No. 1, NASA Contract No. NASr-73.
McIntosh, J. R. (1963). Diffusion, permeation and solution of selected gases in silicone rubber. M. S. thesis, State University of Iowa (1962).
Roberts, R. W. (1962). Effects of polymeric structure upon gas permeation. Ph.D. dissertation, State University of Iowa.
Weller, S. W., and W. A. Steiner (1950). J. Appl. Phys. 21: 279-83.

7.

NEW APPROACHES TO CONTAMINANT CONTROL IN SPACECRAFT

ERIC E. AUERBACH · SID RUSSELL

Hamilton Standard, Division of United Aircraft Corporation
Windsor Locks, Connecticut

INTRODUCTION

As space missions grow longer, the problem of control of trace contaminants to prevent gradual accumulation to harmful concentrations becomes more acute. This specific problem is not unique with spacecraft; it has been prominent, for instance, in submarines for a number of years. In spacecraft we must bring to bear experience gained in the conditioning of submarine atmospheres; however, we must impose stricter limitations on the weight, size, and power of control systems. In addition, we must consider those other aspects, such as zero gravity, subatmospheric pressure, or specific contaminants, which are peculiar to spacecraft.

The successful design of a system to provide spacecraft contaminant control requires that satisfactory answers be provided to four basic questions: (1) What are the contaminants to be considered? (2) What are their generation rates? (3) What are their maximum allowable concentrations? and (4) How can each gas be kept below its maximum allowable concentration? Answers to questions 1 and 2 must rely heavily upon space flight, simulated space flight, and submarine experience. Question 3 is principally one of toxicology. Its solution lies in successful modification of established industrial concentration limits to compensate for such factors as continuous exposure, psychological stress, zero gravity, high activity and metabolic rates, and high oxygen concentration. The final question (4) is one of applied chemistry and engineering, and experimental efforts have been concentrated in this area.

This paper reviews some of the investigations that have been and are being made at Hamilton Standard Division of United Aircraft to define trace contaminant control requirements in a quantitative fashion, and to develop control techniques and devices that may readily be adapted to widely divergent mission requirements. This involves emphasis on a continuing study of physical and chemical sorbents and of combustion catalysts. As a result of these investigations, prototype components have been developed and are being integrated into an all-purpose contaminant control system.

METHODS OF CONTROL

In the Project Mercury and Gemini space flights, toxic gas concentrations have been controlled by (1) leakage, (2) lithium hydroxide, and (3) activated charcoal. It can be shown mathematically that when a steady state has been achieved, and the rate of contaminant disposal is equal to the rate of contaminant generation, the cabin has accumulated a contaminant concentration given in parts per million by

$$\text{p.p.m.} = (10^6)(R_G/R_L)(M_M/M_C)$$

where R_G is the contaminant generation rate in pounds per hour, R_L the rate of gas mixture leakage to space, M_M the mixture molelar weight, and M_C the contaminant molecular weight. Thus, when leakage rate is sufficiently high with respect to contaminant generation rate, a low concentration of contaminant is sustained.

Although the primary purpose of the lithium hydroxide is to remove the carbon dioxide produced within the spacecraft, it is also effective in absorbing certain acid gases present in trace quantities. In operation, the cabin atmosphere is circulated through a bed of solid, anhydrous lithium hydroxide granules, where the carbon dioxide is chemisorbed:

$$CO_2 + 2LiOH = Li_2CO_3 + H_2O$$

Lithium hydroxide has a capacity in most applications approaching the stoichiometric value of 0.917 lb. of carbon dioxide per pound of lithium hydroxide. Once its capacity has been attained, however, it cannot be regenerated practicably.

Activated charcoal is currently depended on to remove those toxic gases not removed by either leakage or lithium hydroxide. Many such contaminants are removed by physical adsorption on the extended surface of the charcoal. The capacity for each contaminant is strongly dependent on the conditions of adsorption.

Deficiencies of current control methods

As the duration of space missions increases, the deficiencies of the foregoing methods of contaminant control become readily apparent. Leakage of the spacecraft atmosphere to space must be made up from gas storage and supply equipment. Thus, for longer missions, storage and supply weight becomes excessive unless leakage is made negligible. Similarly, lithium hydroxide weight becomes excessive for longer missions. Consequently, its use on long duration space flights will be reserved for emergencies. The shortcomings of charcoal, on the other hand, do not arise directly from a weight problem. Although charcoal is invaluable for removing many contaminants, it has very low capacity for low molecular weight gases such as methane, carbon monoxide, and ammonia. Because lithium hydroxide also has low or negligible capacity for such contaminants, additional means of control must be provided.

New approaches to control

To meet the requirements of extended space missions, present contaminant control equipment will have to be supplemented or supplanted by other means for contaminant control. Important among these are catalytic burners, broad spectrum chemisorbent cartridges, and regenerable carbon dioxide removal systems.

Carbon dioxide may be treated separately from other atmospheric contaminants for a number of reasons. While basically similar to many other contaminants, carbon dioxide is unique in that: (1) it is present in concentrations of well over 1,000 p.p.m, and is therefore not a trace contaminant; (2) it is generated at well-defined rates; and (3) methods for its control are well developed and removal equipment may be quantitatively designed with confidence. A variety of systems for carbon dioxide control have been developed or are the object of current research. Many of these form part of a system designed to process the carbon dioxide in order to recover oxygen. Recent developments in regenerable carbon dioxide removal systems are currently receiving particular attention. Control of carbon dioxide has been developed to such a degree and is so extensively reported that it will not be further considered here.

For the remaining diverse trace contaminants, it is the approach of current researchers to discover most efficient removal techniques, to assemble reliable removal performance data, and to provide a sound procedural model upon which developed system components may be selected, sized, and integrated according to varying mission requirements. Fundamental also to a computational

TABLE 1

PARTIAL LIST OF CONTAMINANTS IDENTIFIED IN SEALED ATMOSPHERES MERCURY SPACE FLIGHTS

(Collected on Activated Charcoal)

Contaminant	Formula	Conc. (p.p.m.)[b]	Contaminant	Formula	Conc. (p.p.m.)[b]
1. Freon-114[a]	$CF_2Cl—CF_2Cl$	60-6000	23. Methyl alcohol[a]	CH_3OH	0-1
2. Ethylene dichloride	$CH_2Cl—CH_2Cl$	0-40	24. 1,4-Dioxane[a]	$(CH_2)_4O_2$	0-1
3. Toluene[a]	$C_6H_5CH_3$	3-20	25. Cyclohexane[a]	$(CH_2)_6$	0-1
4. n-Butyl alcohol	C_4H_9OH	0-4	26. Formaldehyde	CH_2O	—
5. Freon-11	$CFCl_3$	0-3	27. Hexamethylcyclo-trisiloxane	$(CH_3)_6(SiO)_3$	—
6. Vinyl chloride	$CH_2=CHCl$	0-3	28. Freon-22	CHF_2Cl	—
7. Ethyl alcohol[a]	C_2H_5OH	0-3	29. Freon-23	CHF_3	—
8. m-Xylene	$C_6H_4(CH_3)_2$	0-3	30. Freon-12	CF_2CCl_2	—
9. Vinylidene chloride[a]	$CH_2=CCl_2$	0-2	31. Freon-125	CF_3CF_2H	—
10. Methylene chloride[a]	CH_2Cl_2	0-2	32. Hexene-1	C_6H_{12}	—
11. o-Xylene	$C_6H_4(CH_3)_2$	0-1	33. Propylene	C_3H_6	—
12. Benzene[a]	C_6H_6	0-1	34. n-Butane	C_4H_{10}	—
13. Methylchloroform	CH_3CCl_3	0-1	35. Butene-1	C_4H_8	—
14. Trichloroethylene	$CHCl=CCl_2$	0-1	36. iso-Pentane	C_5H_{12}	—
15. Acetone	CH_3COCH_3	0-1	37. n-Pentane	C_5H_{12}	—
16. Methyl ethyl ketone	$CH_3COC_2H_5$	0-1	38. Propane	C_3H_8	—
17. Methyl isopropyl ketone	$CH_3COC_3H_7$	0-1	39. n-Hexane	C_6H_{14}	—
18. Ethylene	$CH_2=CH_2$	0-1	40. 2,2-Dimethylbutane	C_6H_{14}	—

Contaminant	Formula	
19. n-Propyl alcohol	C_3H_7OH	0–1
20. Acetaldehyde	CH_3CHO	0–1
21. Ethyl acetate	$CH_3COOC_2H_5$	0–1
22. Freon–114 (unsym.)	$CFCl_2CF_3$	0–1
41. $trans$-Butene–2	C_4H_8	—
42. cis-Butene–2	C_4H_8	—
43. Acetylene	C_2H_2	—
44. 3-Methylpentane	C_6H_{14}	—

Nuclear submarines

Contaminant	Formula
45. Carbon monoxide	CO
46. Hydrogen	H_2
47. Methane	CH_4
48. Chlorine	Cl_2
49. Hydrogen fluoride	HF
50. Ammonia	NH_3
51. Nitrogen dioxide	HO_2
52. Stibine	SbH_3
53. Dimethyl-5 ethyl benzene	$(CH_3)_2C_6H_3C_2H_5$
54. p-Ethyl toluene	$C_2H_5C_6H_4CH_3$
55. Gasoline vapors	
56. Mesitylene	$C_6H_3(CH_3)_3$
57. Pseudocumene	$C_6H_3(CH_3)_3$
58. p-Xylene	$C_6H_4(CH_3)_2$
59. Hydrogen chloride	HCl
60. Sulfur dioxide	SO_2
61. Arsine	AsH_3

[a] These contaminants were common to the atmospheres of the first three U.S. manned orbital flights.

[b] The values listed represent the approximate minimum concentrations that would have ensued had all of the recovered contaminant been dispersed in the free volume of the cabin at one time. A dash indicates that quantitative values were not determined.

approach to control system construction, and the further object of concurrent research, is the ability to precisely and quantitatively define control requirements, i.e., contaminants, generation rates, maximum allowable concentrations, and other pertinent system operating parameters and limits.

ESTABLISHING TRACE CONTAMINANT CONTROL REQUIREMENTS

Although nuclear submarines have been the major source of contaminant data for closed systems, there is a growing body of information from manned space flights and from simulated flights. Table 1 presents a list of some of these contaminants and illustrates the variety found. These contaminants are emitted by many different types of materials, resulting in uncertain predictability. Man, however, is always present, and his bioeffluents would appear simple to predict. Indeed, carbon monoxide and even methane and hydrogen sulfide are generated at reasonably predictable rates. Nevertheless, some metabolic contaminants present greater difficulties. For example, lactic acid generation rate depends upon degree of physical activity, but its resulting atmospheric concentration depends upon its transport rate from liquid perspiration to the atmosphere, and the amount of perspiration depends not only on the degree of physical activity, but also on the means of thermal control used in the spacecraft.

Examples of other contaminants that have been found in closed environmental systems are: ammonia from waste disposal systems; solvents from paints and plastics; alcohol from equipment sterilization processes; Freon vapor from leaks in heat transfer equipment; and irritating odors from overheated electric motor lubricants.

Contaminant generation rates

Generation rate determines first, the required process air flow for a given maximum allowable concentration, and second, the quantity of material (for example, charcoal) needed to control the contaminant for a given period of time. Table 2 lists some generation rates and maximum allowable concentrations for spacecraft that are serving current design efforts. They are based on current requirements for a specific two-man mission as indicated by prior closed system study.

The need for tempered application of such quantitative generation rate data cannot be overemphasized. Even though the composi-

TABLE 2

TRACE CONTAMINANT GENERATION RATES AND MAXIMUM
ALLOWABLE CONCENTRATIONS FOR A SPECIFIC TWO-MAN MISSION

	Estimated generation rate (lb./hr. $\times$ 10^8)	Space maximum allowable conc. (p.p.m.)
Ammonia	8,300	10
Benzene	830	5
Carbon monoxide	230	20
Cyclohexane	170	80
Dioxane	83	20
Ethanol	830	200
Formaldehyde	83	1
Hydrogen	1,640	41,000
Hydrogen fluoride	250	1
Hydrogen sulfide	3.58	2
Methane	144,000	200
Methanol	830	40
Methylene chloride	83	100
Ozone	8.3	0.02
Sulfur dioxide	83	1
Toluene	830	40

tion and total quantities of respired air, flatus, and sweat are fairly well established, contamination rates from these sources are often difficult to predict with great certainty for different space capsule configurations. Likewise, contaminant generation rates from the the immediate environment of the crew are not so easily defined, owing to the variability of the sources, the lack of extensive test data, and the ever-present possibility of equipment malfunction. Programs to implement these data by identifying and determining generation rates of contaminants from a wide assortment of materials of construction, which will ultimately aid in materials selection, are now in progress at Hamilton Standard and other interested companies.

Thus, generation rate data that may be applied directly to any selected mission cannot be predicted with great accuracy until flight or simulated flight data with the spacecraft configuration in question become available. Nevertheless, data (such as those in Table 2) that reflect accumulated information are useful today and are undergoing constant improvement as new findings become available. At the same time, reduction in the number of contaminants and in generation rates is being accomplished by a more careful selection of materials.

Maximum allowable concentrations

The threshold limit values (TLV) that have been established by the American Conference of Government Industrial Hygienists (1964), pertain to an 8-hour daily exposure and are not directly applicable to a space mission. Stokinger (1963) pointed out that these standard TLV's must be modified not only with respect to continuous exposure, but also total pressure, ambient temperature, movement restriction, oxygen concentration, fatigue, and contaminant interaction. For space application, the space maximum allowable concentration (SMAC) is generally set by decreasing the TLV by a factor of 10 or less, although the modification can be more drastic. For example, Stokinger recommends (for a 90-day mission in pure, 5 p.s.i.a. oxygen) that the industrial TLV of 1 p.p.m. for hydrazine be reduced to 0.019 p.p.m., a factor of 52.5.

In addition to the foregoing modifications, the influence of odor and additive physiological action must be considered. The industrial TLV's, as well as their modifications, are not based on unpleasant odor. Nevertheless, it is a general aerospace requirement that unpleasant odors be prevented. Odor thresholds are, in fact, usually lower than maximum allowable concentrations based on toxicity, so that many of the modified TLV's may not be the determining criteria.

Modified TLV's that are pertinent must be further corrected for additive physiological action. Contaminants that attack the same body organ may be said to comprise a toxicological group. If the members of a given group are expected to be present simultaneously, it is logical that the individual maximum allowable concentrations should be reduced such that

$$\sum_{i=1}^{n} \left(\frac{C_i}{SMAC_i} \right) = 1$$

where i is any one of a toxicological group of n contaminants, $SMAC$ the individual maximum allowable concentration based on toxicity, and C the maximum allowable concentration of each contaminant when in the mixture (Am. Conf. Gov. Ind. Hyg., 1964). For example, suppose the individual maximum allowable concentrations are 20 p.p.m. for toluene and 40 p.p.m. for cyclohexane, both of which may cause narcosis. If these two contaminants, when present at the same time, are limited to concentrations of

15 p.p.m. for toluene and 10 p.p.m. for cyclohexane,

$$\sum_{i=1}^{n} \left(\frac{C_i}{SMAC_i} \right) = \frac{15}{20} + \frac{10}{40} = 1$$

and the limiting concentrations are satisfactory. Nevertheless, the complexity of the situation is further emphasized by the fact that a single contaminant may belong to two different physiological groups and must be treated as a member of each. Application of the foregoing considerations for purposes of design is considered later.

METHODS FOR CONTROL OF TRACE CONTAMINANTS

In general, all trace contaminants can be controlled or removed in one or a combination of the following ways: (1) filtration; (2) physical adsorption; (3) chemical absorption; (4) chemical or catalytic conversion to nontoxic end products; (5) conversion to a more easily controllable contaminant.

Leakage, as discussed earlier, is generally unsatisfactory as a means of control. Determination of a specific control method for each contaminant that might find its way into a manned space vehicle would involve a task of almost unlimited magnitude. However, the problem of contaminant control may be attacked in much the same manner as many other chemistry problems—by classifying the contaminants into groups reflecting their general chemical and physical similarities. Thus, there are hydrocarbons (which are completely oxidizable to carbon dioxide and water); the acid gases such as hydrogen sulfide, chlorine, and nitrogen and sulfur oxides; the alkaline gases such as ammonia and volatile amines; halogenated hydrocarbons; and so on. In this manner, it is possible to organize all contaminants into five general groups. By optimizing control processes for representative or critical compounds from each group, it is in general possible to optimize control techniques for the entire group. The five groupings of contaminants selected by Hamilton Standard are shown in Table 3, where the entire listing of Table 1 has been reorganized accordingly. As new contaminants are recognized, they are incorporated into one of these groups.

The high molecular weight contaminants of Groups I through III can generally be physically adsorbed on such materials as activated charcoal. Lower molecular weight constituents of Group I cannot be readily adsorbed, but are completely oxidizable to

TABLE 3

CLASSIFICATION OF CONTAMINANTS LISTED IN TABLE 1

Group I C, H, O compounds	Group II Acid or alkaline gases	Group III Near-neutral gases: halogen, nitrogen, sulfur, heavy metal, etc., compounds	Group IV dusts and smoke	Group V others
1. Toluene	40. Chlorine	46. Freon-114	Aerosols	Aerosols
2. n-Butyl alcohol	41. Hydrogen fluoride	47. Ethylene dichloride		Bacteria
3. Ethyl alcohol	42. Ammonia	48. Freon-11		Colon-inhabiting
4. n-Xylene	43. Nitrogen dioxide	49. Vinyl chloride		bacilli
5. o-Xylene	44. Hydrogen chloride	50. Vinylidene chloride		Diplococci
6. Benzene	45. Sulfur dioxide	51. Methylene chloride		Spore-forming
7. Acetone		52. Methyl chloroform		bacilli
8. Methyl ethyl ketone		53. Trichloroethylene		Staphylococci
9. Methyl isopropyl ketone		54. Freon-114 (unsym.)		Streptococci
10. Ethylene		55. Hexamethylcyclotrisiloxane		Ions
11. n-Propyl alcohol		56. Freon-22		
12. Acetaldehyde		57. Freon-23		
13. Ethyl acetate		58. Freon-12		
14. Methyl alcohol		59. Freon-125		
15. 1, 4-Dioxane		60. Stibine		
16. Cyclohexane		61. Arsine		

Group I also includes: 17. formaldehyde; 18. hexene-1; 19. propylene; 20. n-butane; 21. butene-1; 22. iso-pentane; 23. n-pentane; 24. propane; 25. n-hexane; 26. 2, 2-dimethylbutane; 27. trans -butene-2; 28. cis-butene-2; 29. acetylene; 30. 3-methylpentane; 31. carbon monoxide; 32. hydrogen; 33. methane; 34. dimethyl-5 ethyl benzene; 35. p-ethyl toluene; 36. gasoline vapors; 37. mesitylene; 38. pseudocumene; 39. p-xylene.

carbon dioxide and water. Among these are hydrogen, carbon monoxide, methane, and other hydrocarbons that can be oxidized in a catalytic burner. Lower molecular weight constituents of Groups II and III that cannot be physically adsorbed on charcoal would be converted to other equally harmful contaminants in a catalytic burner (such as halogen, nitrogen, and sulfur oxides), so that they must be controlled by other means, although chiefly by chemisorption. These three methods (physical adsorption, catalytic combustion, and chemisorption) provide primary control of gaseous contaminants, Groups I through III. Group IV, particulate matter, is controlled by appropriate filtration. The extent of control required for Group V, aerosols and bacteria, is presently undefined; however, most of these contaminants can be controlled effectively by one or more of the processes just described. Thus, aerosols can be removed in the filtration, oxidation, and sorption processes. Most bacteria will also be destroyed during these same processes, although they may require an additional control source, such as ultraviolet radiation. Ions may require electrostatic precipitation or neutralization, if they are not readily captured by the aerosols present.

An experimental program to investigate and further develop methods of trace contaminant control was initiated at Hamilton Standard in 1961. This program has sought to evaluate, select, and develop improved catalysts, chemical sorbent materials, and charcoal types for contaminant control, and to develop the analytical tools necessary to indicate contaminant control device performance, at the pertinent trace concentration levels. Maintenance of effectiveness over a wide range of humidity in the presence of 0.5% carbon dioxide is an additional requirement for these devices.

CATALYTIC COMBUSTION

Oxide and precious metal catalysts on assorted supports have been evaluated for effectiveness to convert oxidizable air contaminants to products that can be controlled with current carbon dioxide and humidity control systems. Oxidation efficiency data have been developed as a function of temperature, flow rate, humidity, and catalyst bed length at trace concentration levels. This has been accomplished through a research program and a prototype test program.

Research program

The research program sought to establish the preferred oxidation catalysts and to indicate the influence of the various operating parameters on conversion efficiency. The principal contaminants studied were hydrogen, carbon monoxide, and methane, all of which were known to be produced metabolically. Ultimately, the criteria for catalyst selection were (1) minimum pressure drop, and (2) minimum oxidation temperature for methane, the contaminant requiring the highest oxidation temperature.

The experimental equipment consisted of a gas supply, devices for control of test conditions, a catalytic reactor, and measuring instruments. The laboratory setup is represented in Figure 1. A calibrated mixture of the contaminant in dry air was supplied from a high pressure cylinder. Cylinders containing various concentrations of contaminant were prepared by charging a cylinder of pure contaminant at low pressure with high pressure air. Control of operating parameters was straightforward: flow rate by valves, humidity by a bubbler with bypass, and temperature by a heating mantle surrounding the reactor. The reaction chamber was a 2-in.-long stainless steel tube, containing the catalyst held between two glass wool plugs. Test conditions were measured with a rotameter, a pressure gage, an Alnor Dew Pointer, and a thermocouple inserted axially through the reactor outlet tube into the glass wool plug.

Concentrations of contaminants were measured with a Jarrell-Ash Universal Gas Chromatograph, equipped with an argon ionization detector. Gas samples were obtained by syringe through rubber septa at the reactor inlet and outlet. Once the sample was injected into the chromatograph, a strip chart giving the results of the analysis was obtained directly. Although it was necessary to use two different chromatographic columns to determine carbon monoxide and carbon dioxide, simultaneous measurement, with some loss in sensitivity, was obtained by use of parallel columns. When hydrogen and carbon monoxide were catalytically oxidized below 150°F., methane was unreactive and was therefore useful as a tag gas to validate the observed carbon monoxide and hydrogen concentration changes from reactor inlet to outlet. Techniques used to measure concentrations of less than 1 p.p.m. with an ionization detector and helium carrier gas were described in detail in a previous paper by Russell (1964).

Results of this experimental work indicated the temperatures required for complete oxidation of the contaminants and also the effect of the several operating parameters on oxidation efficiency utilizing a variety of catalysts. Without exception, the effect of increasing contaminant inlet concentration was to decrease oxidation efficiency, although for methane this trend was negligible. Increasing the flow rate, and therefore decreasing the contact time, always resulted in decreased efficiency, this decrease being

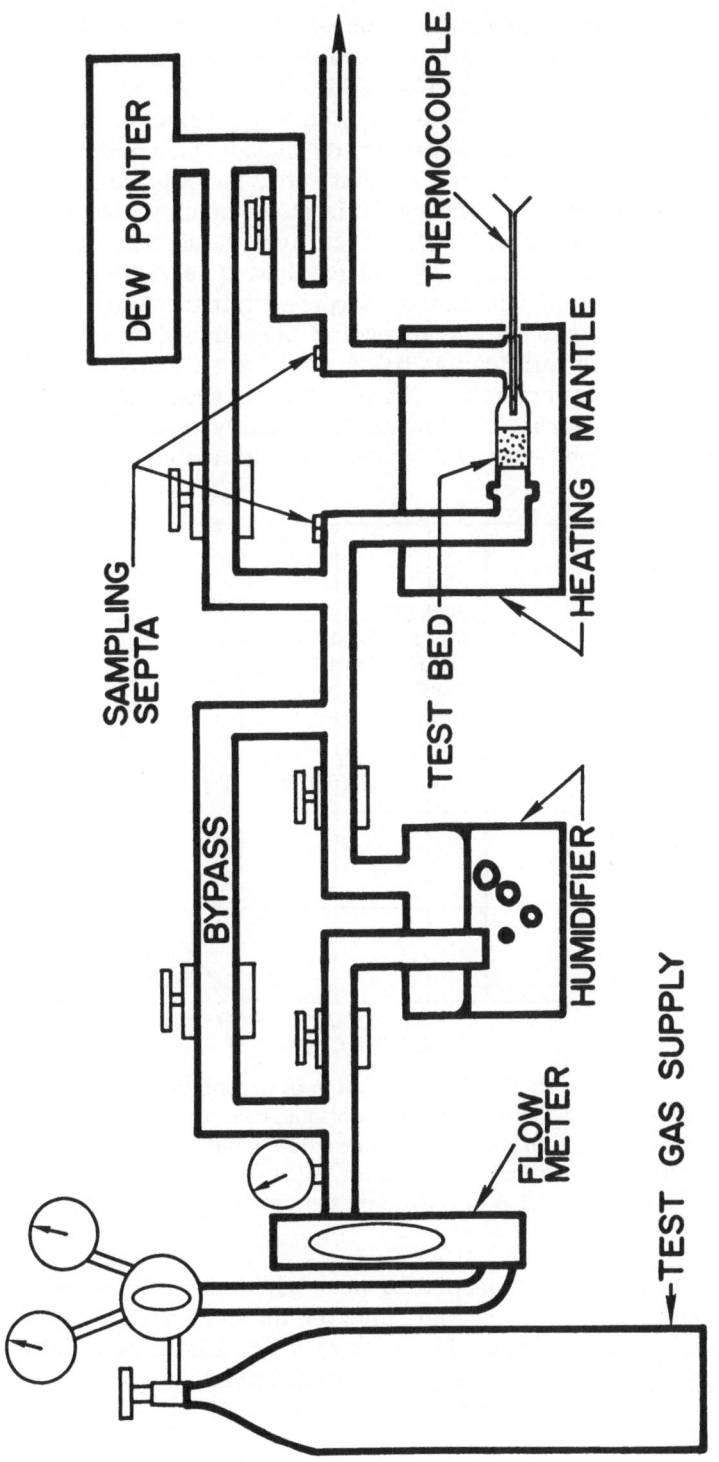

Figure 1. Catalytic oxidation test apparatus.

sharper at lower temperatures. Increasing the length of the catalyst bed, of course, effected a considerable increase in efficiency, except at efficiencies approaching 100%. Increasing reaction temperature always increased efficiency. Methane required a much higher temperature than either hydrogen or carbon monoxide, as shown in Figure 2. Inlet moisture content often had a pronounced effect on reaction efficiency, although this effect quickly disappeared as temperature was raised above a critical value that depended on such factors as contaminant and catalyst type. Efficiency never increased but often decreased sharply with increasing inlet moisture content. Some of the foregoing phenomena, which relate to general trends at efficiencies below 100%, are illustrated in Figure 3. Additional experiments showed that carbon dioxide, always present in spacecraft atmospheres, probably has no effect on the oxidation reactions when it is present in anticipated concentrations.

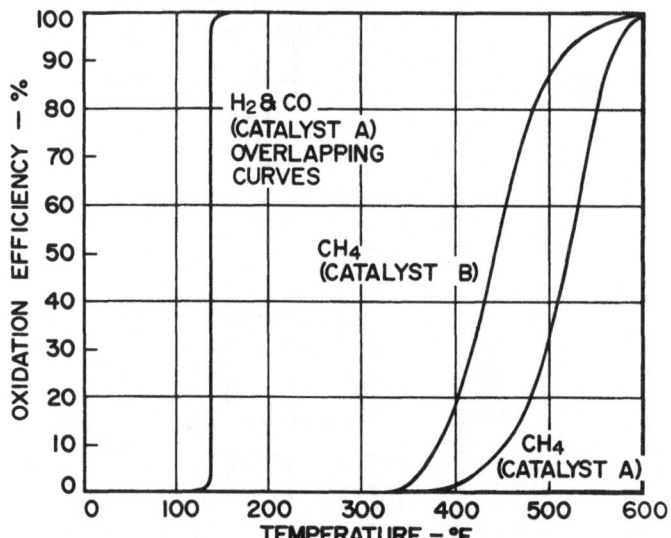

Figure 2. Catalytic oxidation of carbon monoxide, hydrogen, and methane. Test conditions: concentration of each contaminant in air, ~ 75 p.p.m.; dew point, ~ +15° F.; bed length, 1 in.; bed O.D., 0.5 in.; flow rate, 280 cc./min. Catalysts A and B are platinum family metals deposited on a metal oxide carrier.

The experiments described here yielded the expected effects of operating parameters and indicated the preferred catalysts and the operating conditions they required. The basis for development of a prototype reactor was thereby established.

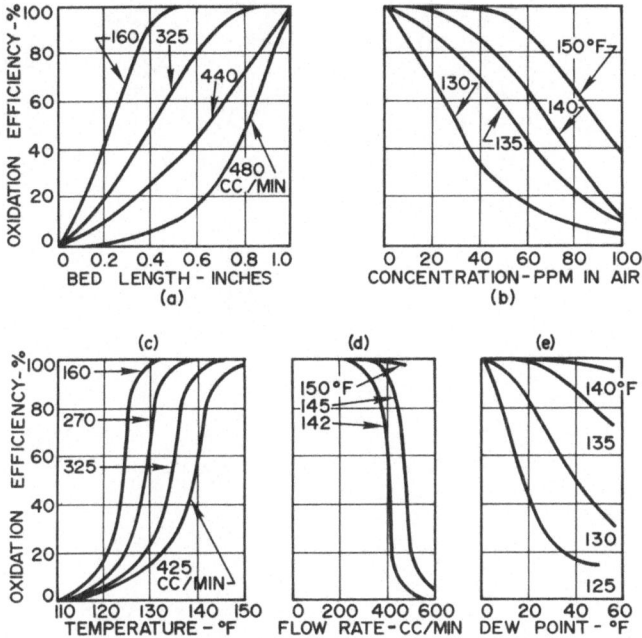

Figure 3.   Carbon monoxide oxidation. (Catalyst used was palladium on asbestos.)

Prototype development program

Based on results from the research laboratory catalytic combustion studies, a flight prototype catalytic burner was designed and tested. This device, on which a patent is pending, not only incorporated the most efficient catalyst, but also included new concepts of reducing operating power requirements.

This prototype catalytic burner is shown in Figure 4. In operation, the coiled regenerative heat exchanger recovers heat by preheating the inlet air with the hot exit air. The inlet air is carried the length of the catalyst chamber in an axial tube and impinges on the heater, which forms one end of the chamber. This hot air is then reflected through the catalyst bed, and passes out through the regenerative heat exchanger. Radiative and convective heat losses are minimized by use of a superinsulation. Even at an interior temperature of 1,000°F., the exterior casing is safe to touch, although the design operating temperature is only 550°F.

The prototype test setup was in some ways similar to that used in the research work. The major differences were (1) provision to operate at subatmospheric pressure; (2) provision to evacuate superinsulation; (3) use of a self-contained reactor heater; and (4)

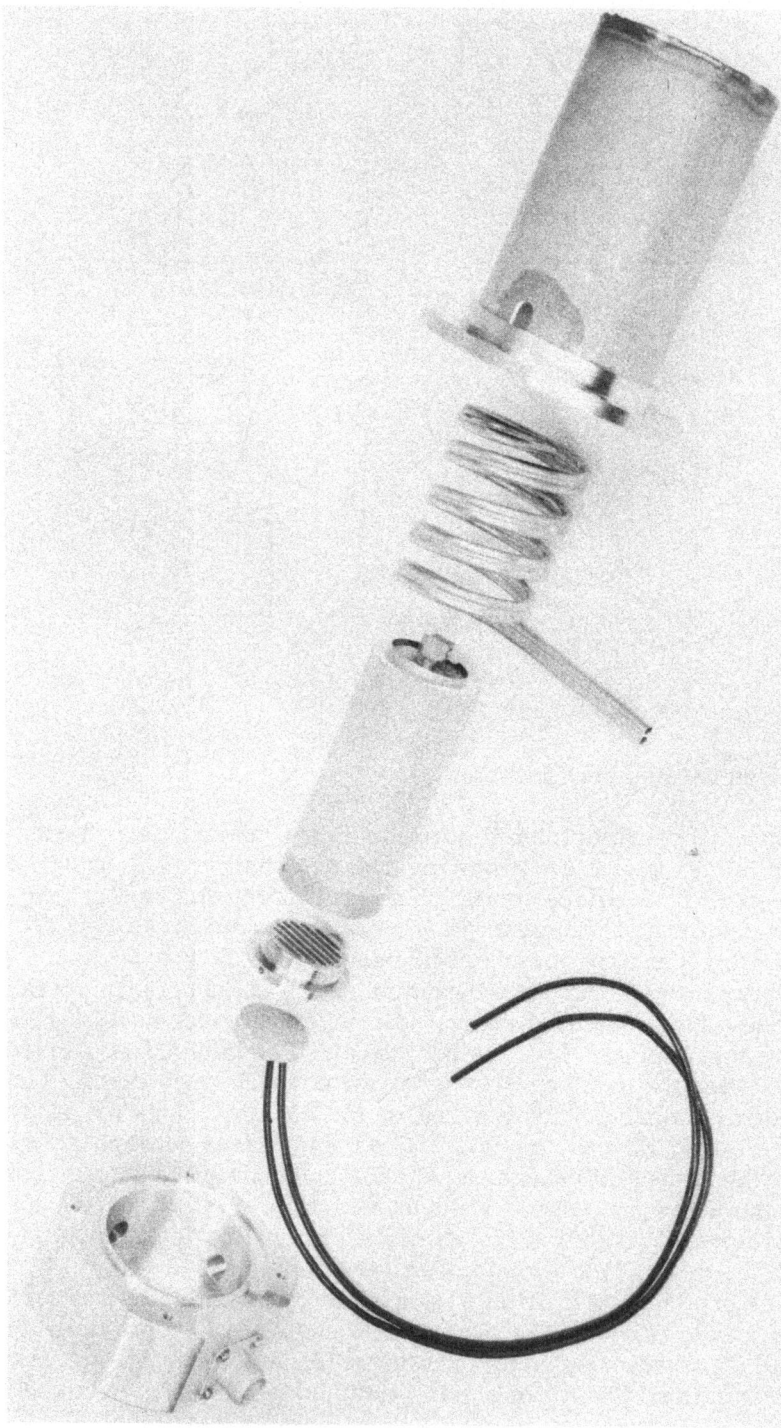

Figure 4.   Prototype catalytic burner (patent pending).

use of a single gas supply cylinder, containing a mixture of trace contaminants in oxygen. In addition to oxygen, the supply cylinder contained hydrogen, carbon monoxide, methane, and carbon dioxide. Control of the operating conditions was manual except for the pressure regulators, and much of the work was done at a pressure of 5 p.s.i.a. Performance of the superinsulation was studied at various pressures, as low as 0.1μ, maintained with an oil diffusion pump.

Instrumentation was generally more extensive than that used in the research work. The same gas chromatograph measured contaminant concentrations. Operating pressures were measured with manometers, and insulation pressure (at the outermost layer) was measured with thermal and ionization gages. Temperatures were measured at several points within the reactor and heat exchanger.

Results for the effects of temperature and flow rate were similar to those obtained in the research work. Data for operating power required, however, were entirely new; oxidation tests were carried out at temperatures from 250°F. to 600°F. At these temperatures, carbon monoxide and hydrogen were completely oxidized. Oxidation of methane was represented by characteristic S-curves of almost exactly the same shape as obtained in the research experiments, as shown in Figure 5. As indicated by Figure 5, reaction temperature required for a specified oxidation efficiency increased with increasing total mass flow rate, but this effect was less evident at higher flow rates. Power required was found to be very much dependent on insulation pressure above a value of 5μ (because of gas conduction within the insulation), but became independent of insulation pressure in the 0.1-μ region. Power required was also extremely dependent on reaction temperature, because of associated heat loss by conduction through metallic parts and by radiation through the insulation. Required power also rises as total flow rate is increased, because of sensible heat loss in the gas stream. The dependence of total operating power on flow rate and reaction temperature is shown in Figure 6.

The development tests of the prototype catalytic burner showed that a safe, compact, lightweight reactor is practical for use on spacecraft. Fixed weight and pressure drop are reasonable, while required power of the device is exceptionally low.

CHEMICAL AND PHYSICAL SORPTION

Many trace contaminants that cannot be controlled by either adsorption on activated charcoal or oxidation to less toxic products

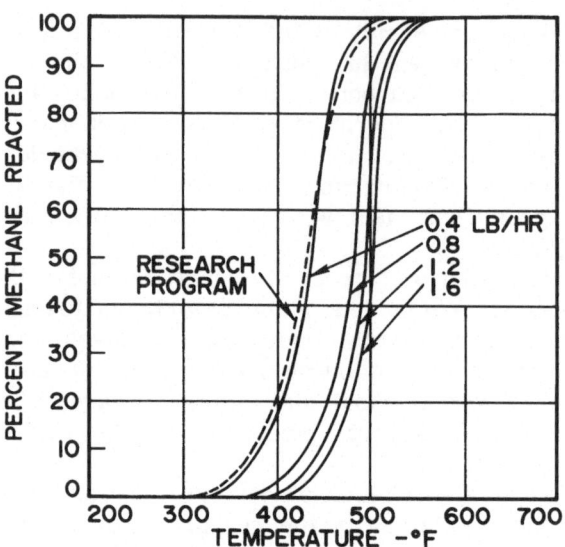

Figure 5. Prototype catalytic burner performance (atmospheric pressure).

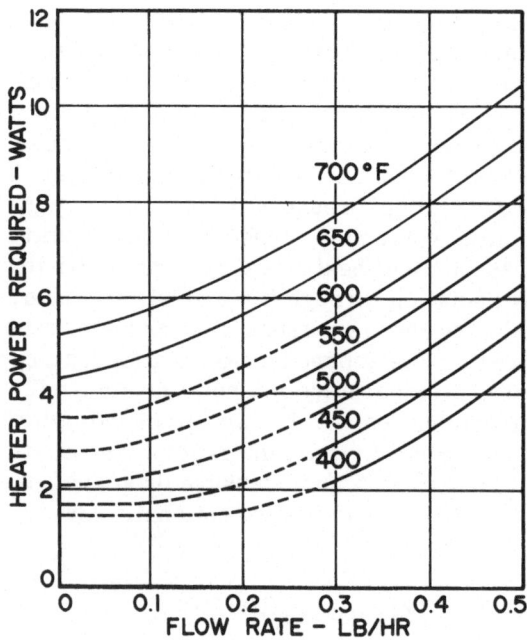

Figure 6. Catalytic burner power required.

in a catalytic burner may be removed from the spacecraft atmosphere by sorption on other solid adsorbents or by reaction on solid chemicals. These contaminants constitute Groups II and III in Table 3. They include acid gases such as hydrogen sulfide; alkaline gases such as ammonia and low molecular weight amines; neutral vapors, including low molecular weight halogenated hydrocarbons; and other vapors, including mercaptans, organic metal compounds such as carbonyls, and inorganic metallics such as mercury vapor. Research has been conducted to select sorbents for such contaminants, leading to development of a prototype sorbent canister.

Sorption research studies

The goal of our continuing sorption research program is to select chemicals that will remove gaseous contaminants that cannot be controlled by oxidation, and to learn how various operating conditions affect sorbent performance. In addition, the interaction of common spacecraft materials such as charcoal, lithium hydroxide, and molecular sieves (from regenerable solid adsorbent systems) with these contaminants is being determined.

The general pattern of investigation for each contaminant consists of the following steps: (1) development of suitable analytical techniques for detection and measurement of the contaminant at very low concentrations; (2) screening of candidate sorbents by determining capacity at contaminant breakthrough; and (3) comprehensive testing of the most promising sorbents under a variety of conditions.

The experimental equipment is much the same as that used for the catalytic combustion work. The contaminant-air mixture, which is supplied either from a prepared cylinder or a gas generator, is humidified before passing through the sorbent material. Presence of the contaminant at the sorbent bed outlet is detected by gas chromatography or, where preferable, by use of color change detection tubes. Use of detection tubes permits determination of sorbent capacity at contaminant breakthrough, the time of first deviation from 100% removal efficiency. The chromatographic techniques permit, in addition, continued determination of sorbent performance where removal efficiencies are below 100%.

Suitable sorbent materials have been obtained for ammonia, hydrogen sulfide, hydrogen chloride, and other contaminants. To be suitable, a sorbent must remove a reasonable quantity of a gaseous contaminant with high efficiency, under the variety of combinations of temperature, humidity, flow rate, and concentration

likely to occur in a spacecraft. Moreover, the sorbent must be effective in the presence of carbon dioxide.

The study of sorption of hydrogen sulfide may be used for
illustration. Difficulty with overloading of the chromatograph ionization detector (since overcome) led to use of MSA detection tubes
as hydrogen sulfide indicators. Comparison of breakthrough time
for a number of possible sorbents, including charcoal and lithium
hydroxide, led to the selection of lead acetate and lead dioxide as
prime candidates. For each of these materials, the effects of inlet
conditions on time to hydrogen sulfide breakthrough (and therefore
on capacity) were determined. These conditions included temperature, flow rate, humidity, contaminant concentration, and the
presence of carbon dioxide. Some of the data collected are shown
in Figure 7. Lead dioxide had higher capacity at all temperatures
studied. It is of interest that sorption of hydrogen sulfide also illustrates the fact that if a sorbent has good capacity, it may be
used, even if the sorption process produces a noxious by-product,
by proper system integration. Thus, absorption of hydrogen sulfide
by lead acetate produces the undesirable by-product acetic acid,
but this acetic acid is destroyed in the catalytic burner. Of course,
the system must be designed so that the catalytic burner flow is
sufficient to control the atmospheric acetic acid concentration to

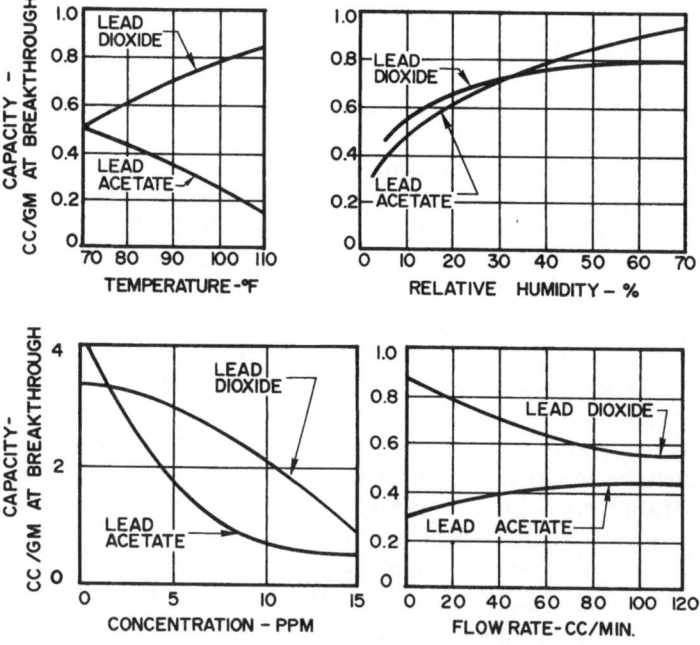

Figure 7. Chemisorption of hydrogen sulfide

an acceptable value. The presence of carbon dioxide had negligible effect on hydrogen sulfide absorption by lead dioxide and lead acetate. In addition, capacity of both lithium hydroxide and activated charcoal was found to be less than one-tenth the capacity of lead dioxide.

Proceeding in a manner similar to that described earlier sorbent materials have been selected for a number of likely spacecraft contaminants, and their individual performance has been evaluated. This is not to say that each contaminant requires a different sorbent. In fact, we have found that similar members of the contaminant groups described earlier are generally removed by the same sorbent material. The information obtained from this research program was sufficient to initiate development testing of a prototype sorbent canister.

Prototype development program

Development of the chemisorbent canister is being carried out in two phases, the first of which is now in progress. The first phase calls for investigation of the performance of a canister containing the various sorbents, when exposed to a mixture of trace contaminants. The second phase involves integration of the canister into a balanced contaminant control system, including particulate filter, activated charcoal, catalytic burner, and carbon dioxide removal components. The goal is to provide a practical system that can control the concentration of any airborne spacecraft contaminant to a safe value.

The test system is flexible. Contaminant gases are supplied from several cylinders, each of which contains several contaminants belonging to the same contaminant group, as described earlier. This permits variation of contaminant concentrations for any group and variation of relative concentrations among groups. The canister itself is sized for use in a typical two-man, 90-day space mission, and is extensively instrumented. It consists of several sorbent compartments, with provision to measure contaminant concentrations not only at inlet and outlet, but also between compartments. The test plan calls first for testing one group of gases at a time, and then all groups simultaneously. With this system built of engineering rather than laboratory hardware, the influence of inlet conditions on sorbent capacity is re-evaluated and, in addition, contaminant interaction is observed. Results of this work are not yet available for publication.

APPLICATION OF RESULTS TO SYSTEM DESIGN

Final design of a spacecraft contaminant control system requires precise definition of the two elements discussed in previous sections: specification of system performance requirements, and generalized correlations of component performance data. The system performance requirements consist of a generation rate and a maximum allowable concentration for each independent contaminant. The component performance correlations must represent the dependence of removal capacity and efficiency on system operating conditions. From this information a contaminant control design procedure may be developed.

Contaminant control design procedure

The purpose of the following step-by-step design procedure is to illustrate a systematic means possible for establishing system flow rates and component arrangement. It is based on the premise that physiological action of the contaminants as a group must be considered, as discussed earlier. Group action may be determined from information on the individual contaminants. The design method is based in part on a calculation presented in the threshold limit values handbook (Am. Conf. Gov. Ind. Hyg., 1964) concerning the treatment of contaminants with similar toxicological action. The resulting procedure is:

1. Arrange contaminants in groups such that all members of a given group exhibit similar toxicological action. These groups will obviously be different from those used in developing control techniques. A given contaminant may appear in more than one group. Where a contaminant's concentration is more severely limited by odor than by toxicity, it should be listed both as a member of a toxicological group and as an individual based on odor.
2. List the specified generation rate and maximum allowable concentration for each contaminant.
3. Rearrange contaminants so that they are cross-tabulated by toxicological group and removal method. Contaminants limited by odor threshold should also be listed individually.
4. For each combination of toxicological group and removal method, calculate and list the quantity

$$W = \sum_{i=1}^{n} (G/CE)$$

where W is the required process rate (lb./hr.), G the contaminant generation rate (lb./hr.), C the contaminant allowable concentration based on toxicity or odor (lb. contaminant/lb. gas), and E the fractional removal efficiency for the removal method indicated.

5. For each removal method (catalytic burner, sorbent bed, etc.), list the *maximum* calculated value of W. These values are then the *minimum* required flow rates for each removal device. Required flow rates for carbon dioxide and water removal should also be noted.

6. Lay out several practical schematic diagrams, based on the list of minimum required flow rates, and select the best one on the basis of low equivalent weight and high reliability.

7. Based on the equipment arrangement obtained in the preceding step, determine the inlet conditions to each contaminant removal device. Then size each of these components, using experimentally derived correlations, on the basis of inlet conditions and total weight of each contaminant generated during the mission.

Some of the foregoing steps probably require additional comment. Step 6 in the procedure, system integration, is best amplified by a brief illustration. An example of tabulated minimum required flow rates accompanies Figure 8, which shows a schematic diagram of a system based on these flow rates. An alternative schematic is indicated by relocating the catalytic burner to a position indicated by the broken lines. This alternative version would probably be preferable if the pressure drop through the catalytic burner were higher than that through the heat exchanger. Determination of the optimum arrangement may not be so simple as minimizing the sum of (1) system equivalent weight due to pressure drop; (2) equivalent weight due to electrical heaters and actuators; and (3) fixed component weight. Indeed, the efficiency of each component depends on its location relative to the other components. For example, placing the catalytic burner so that the flow order is burner-heat exchanger-chemisorbent, may reduce the power requirement of the burner by increasing its inlet temperature, but it may also decrease efficiency of the burner by exposing it to potential catalyst poisons. Thus, true optimization of contaminant control equipment requires thoughtful consideration of the entire system and its interaction with other systems.

Step 7, equipment sizing, requires use of experimentally derived correlations. Such a correlation could, for example, be derived from the information presented in Figure 7, which shows the dependency of sorbent capacity for hydrogen sulfide on temperature, relative humidity, concentration, and flow rate. One method, of limited validity, is multiplication of capacity obtained

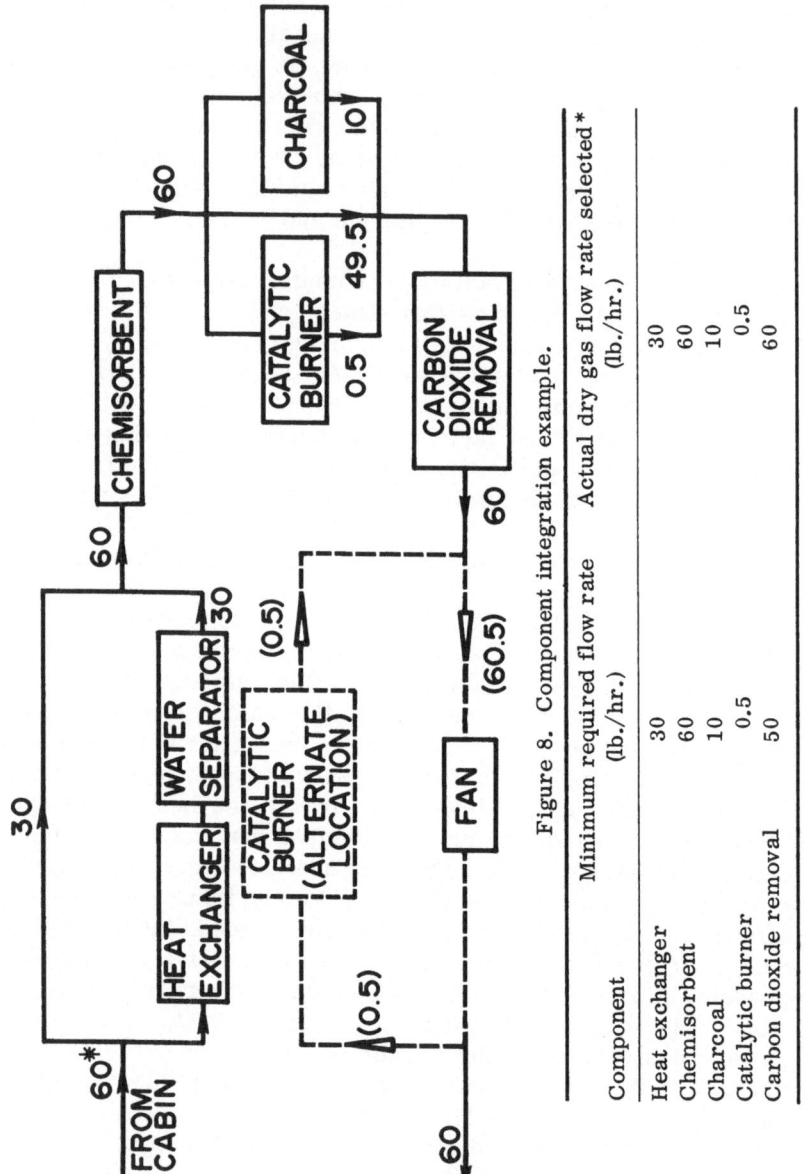

Figure 8. Component integration example.

Component	Minimum required flow rate (lb./hr.)	Actual dry gas flow rate selected* (lb./hr.)
Heat exchanger	30	30
Chemisorbent	60	60
Charcoal	10	10
Catalytic burner	0.5	0.5
Carbon dioxide removal	50	60

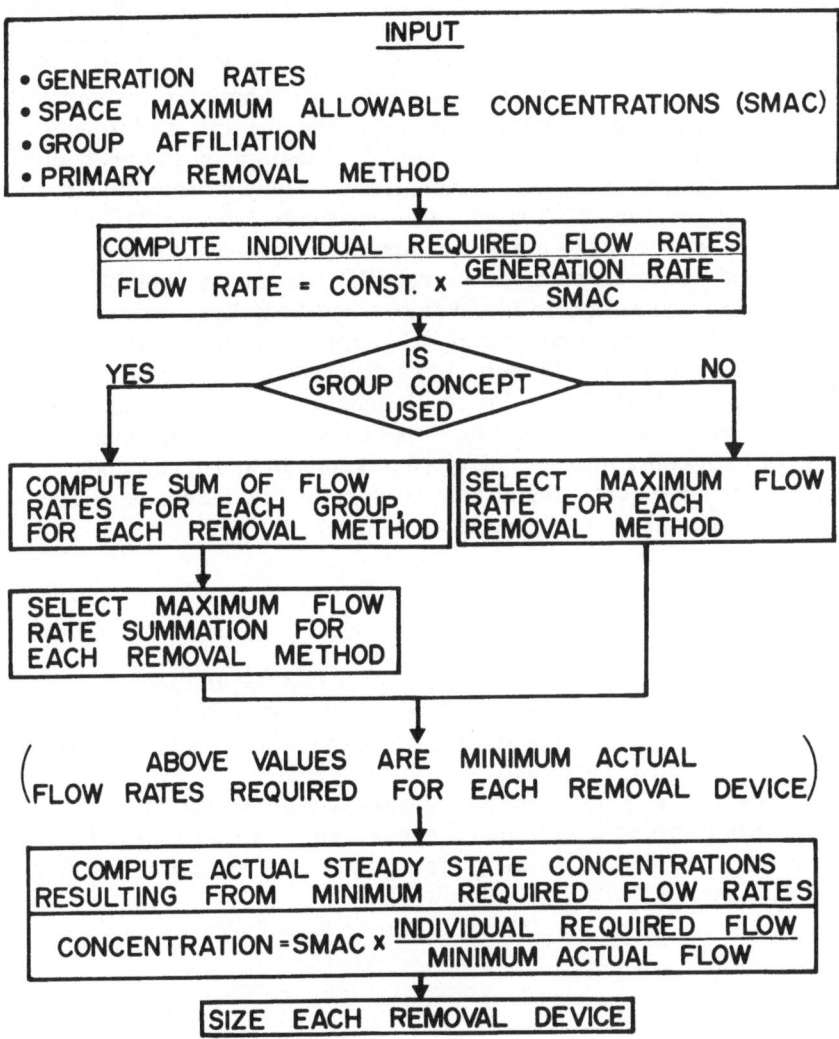

Figure 9. Contaminant control system design computer program.

from the temperature-capacity plot by correction factors for humidity, concentration, and flow rate.

Dividing the contaminants into several toxicological groups, in accordance with the seven-step procedure just outlined, is the correct approach. On occasion, however, it may be expedient to simplify the design problem by not grouping the contaminants at all. This approach amounts to classifying each contaminant into its own private group, and the minimum required flow rate is simply the largest value of the quantity (G/CE). At the opposite extreme, an ultraconservative design is obtained by considering all contaminants as members of the same group. For this approach, the minimum required flow rate is $\Sigma(G/CE)$. Whatever the approach selected, minimum required flow rates may be adjusted to account for system leakage by a simple modification of the procedure.

The design procedure described here may be implemented by use of a computer. The logic for this computer program is indicated in Figure 9. The large number of possible combinations of contaminants and removal methods is readily managed by establishing a matrix of I contaminants and J removal methods. This program provides the following information: (1) minimum required flow rate for each removal device; (2) actual resulting concentration of each contaminant; and (3) size of each contaminant removal device.

As space missions become longer, demand for rigorous definition of both contaminant control requirements and performance of a comprehensive contaminant control system will increase. Planning to meet these demands must be done now.

REFERENCES

American Conference of Government Industrial Hygienists (1964). Threshold limit values for 1964. Arch. Environ. Health, 8: 552.
Russell, S. (1964). Trace analysis of fixed gases by gas chromatography. A.I.H.A. Quarterly, 25 (4): 359-368.
Stokinger, H. E. (1963). Validity and hazards of extrapolating threshold limit values to continuous exposures. A symposium on toxicity in the closed ecological system, Lockheed Missiles and Space Company, pp. 103-123 (July 29-31, 1963).

8.

AN INTEGRATED PROGRAM APPROACH TO THE CONTROL OF SPACE CABIN ATMOSPHERES

J. E. COTTON · T. M. FOSBERG · L. E. MONTEITH*
R. L. OLSON

Aero-Space Division, The Boeing Company, Seattle, Washington

INTRODUCTION

The atmosphere of a manned space system can contain, in addition to the desired components, a number of gaseous and particulate contaminants. These can accumulate to concentrations that are hazardous to occupants. Consequently, methods must be available for limiting contaminants to safe levels.

Two methods of contaminant control are considered. The first deals with minimizing contaminant formation; the second involves techniques for contaminant removal. Contaminant formation can be reduced by judicious choice of materials but it cannot be eliminated. Removal systems can be provided but power and weight penalties require that all such equipment be of optimum design. Therefore, a combination the two methods must be considered in order to provide optimum contaminant management.

The Boeing Company, recognizing the contaminant problem, has developed a comprehensive contaminant management program that is meant to be integrated with the design, manufacture, and test phases of a space vehicle program. Contaminants and release rates, human tolerance levels and control system operating characteristics are determined and used to establish design requirements and specifications. By paralleling vehicle construction, the management program will provide maximum assurance of a habitable final product.

*Present address: Environmental Health Department, University of Washington, Seattle, Washington.

CONTAMINANT MANAGEMENT PROGRAM

The Boeing Company's contaminant management program contains and coordinates all the materials and subsystem activities necessary to provide a habitable environment within a manned spacecraft. Much of the technical information required to organize this program was developed for the Mercury, Gemini, Apollo, and nuclear submarine programs, along with information developed in the

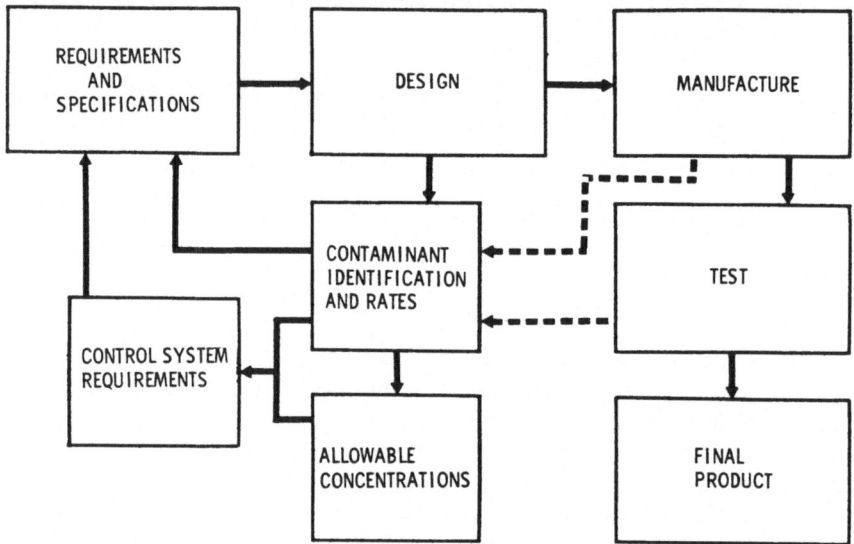

Figure 1. Diagram of contaminant management program.

Minuteman program and in closed-system tests conducted by Boeing.

The program is divided into three basic tasks: (1) identification of contaminants issuing from various sources, with an estimate of their release rates; (2) estimation of allowable contaminant concentrations for extended human exposure; and (3) specification of the control system required to maintain an acceptable atmosphere. The information obtained in these tasks is compiled and made available for use in space vehicle design.

Figure 1 is a simplified diagram of the program showing the relationship of the three tasks in providing design information to the spacecraft system development. The early design is based on the best available, but incomplete, data on contaminants. As additional data become available, successive iterations permit refinements in the design. The object of the program is to establish design requirements and specifications that will provide reasonable assurance that a habitable vehicle will be constructed. Data collected during manufacture and test will insure that this goal is met.

A special feature of our plan has been the establishment of a computer program as the accounting system to calculate and sum the total contaminant production from all sources. An existing weights digital program with detailed breakdown of components and subsystems by material quantity was modified to include contaminant production rate data. The calculations provide an estimate of the concentration of each contaminant in the cabin atmosphere. These estimates may be obtained for any given configuration of a vehicle and for any given mission duration. The results are then compared with man's tolerance to continuous exposure to these contaminants. The excess of potential concentrations over the tolerable limits establishes the removal requirements.

The activities to date have been to conduct analyses utilizing available data for contaminants and production rates, allowable concentrations, and control system requirements. This has allowed assessment of the magnitude of the problem and has identified data deficiencies and problem areas. These will be discussed in following sections.

Contaminant sources

Atmospheric contaminants are evolved from materials of construction and from man. They can also be produced as by-products from the operation of equipment; for example, the contaminant removal sybsystem itself may contribute by converting one contaminant into another.

Contaminants from materials

The first step in control of atmospheric contamination is the detailed evaluation of all materials and subsystems that are proposed for use in a specific vehicle. The evaluation consists of qualitative and quantitative determination of volatile substances produced by each material and subsystem. A detailed accounting of the total amounts of each material used is necessary to permit a prediction of the total release of contaminants.

Metals and inorganic materials used for construction of space vehicles generally present no problem in contamination. A few notable exceptions are cadmium, zinc, mercury, and selenium, for all of which adequate substitutes can be expected to be found. If abrasion (or any other mechanism that will produce microscopic particulate matter) occurs, a problem could arise with many metals or other solid materials.

All organic materials, with only a few exceptions, have been found to produce trace quantities of volatile substances at temperatures of normal usage. It is well known that copious quantities are produced at temperatures only moderately higher than those of normal usage.

The chemical reactions that produce volatile substances can be grouped into three general classifications: (1) diffusion in the solid and vaporization at the surface; (2) oxidation by atmospheric oxygen; and (3) decomposition in the solid followed by diffusion and vaporization.

Many of the volatile products that have been identified can be recognized as solvents and plasticizers used in the manufacture of the organic material. These solvents diffuse to the surface and evaporate at rates determined by the temperature and the time of exposure.

Carbon monoxide has been observed as a product from nearly all organic materials that have been exposed to an atmosphere containing oxygen. This evidently is caused by oxidation at the surface, or, just as likely, by oxidation of vapors in the gas phase. Many aldehydes and organic acids, as well as carbon monoxide, can be accounted for only on the basis of oxidation reactions, since they were known not to be present in the original material.

Thermal decomposition reactions should not take place at significant rates at temperatures of normal usage for most materials. Photochemical decompositions, on the other hand, can be significant if there are radiation sources present.

Volatile substances may be evolved by two or more of the reaction mechanisms cited. It is not necessary to know the chemical reaction mechanism by which each volatile substance arises, but it is helpful. The prime requirement for making the contaminant control program function properly is the compilation of reliable data on the identification and the rate of production of contaminants from all sources. Predicting the contaminants evolved from organic materials constitutes a serious problem, because of the variations in composition of polymers and plastics produced by different manufacturers. There are even significant variations between different batches of a product from the same manufacturer. After the contaminants are determined from a specimen of a given material, therefore, a way must be established to assure that the identical material is used in construction of the vehicle. Two general methods of controlling materials in order to predict contaminants are considered herein.

Source-controlled materials. A very strict material procurement specification is required in this method to carefully control the chemistry of formulation and production of organic materials. The specification is thus concerned with the chemistry of materials and with the identity and rate of production of chemicals resulting from either physical, chemical, or physiocochemical, reactions under conditions of use. It therefore would supplement the functional, physical, or mechanical requirements of material specifications presently imposed on suppliers.

Analysis to determine the identity and amount of each volatile chemical species from each proposed material would be required for the selection and qualification of materials and suppliers. In order to comply with the intentions of this specification, meticulous care would have to be exercised in the formulation, manufacturing processes, and quality control of materials procured.

Batch-tested materials. Materials would be purchased under existing specifications and control would be accomplished by analysis of each batch of materials to determine the identity and amount of each volatile chemical species. A great deal of chemical analysis and careful batch identification procedures would be necessary to assure that the particular batch tested is the material actually used in the vehicle.

Discussions with several prospective suppliers have indicated reluctance on their part to comply with the requirements of the source control method. One very important and valid reason for their reluctance is that proprietary information would frequently be involved. Suppliers cannot be expected to disclose this information, particularly since only relatively small quqntities of these materials would be used. Consequently, the batch-testing method appears more feasible.

Contaminants from humans

Gaseous contaminants are released from the body in the flatus, urine, feces, perspiration, expired air, and eructed air. Under conditions of proper personal hygiene, the escape of some of these substances to the atmosphere, especially those from feces and urine, can be largely prevented. Even with precautions, however, numerous gases will be released from the body.

Contaminants and release rates can be estimated from the known composition and quantity of individual waste products or from data obtained in closed-system tests where total atmospheric contamination from man is measured. Tables presenting the composition of each human waste product have been compiled by Webb (1964). However, it is difficult to determine vaporization rates from sources such as urine, feces, and perspiration since exposure to the atmosphere is variable and decomposition products are not completely known. Man-produced contaminants can be better determined by measurement of accumulation within a sealed chamber. Gorban *et al.* (1964) confined subjects in a 5-cubic-meter airtight chamber for 5 hours. Contaminants released and rates were determined by quantitative analysis of the atmosphere and humidity condensate. The substances identified were ammonia, carbon monoxide, hydrocarbons, aldehydes, ketones, mercaptans, hydrogen sulfide, and fatty acids.

A preliminary estimate of human contaminants released in an operational closed system was made from the data of Gorban *et al.* Compositions of individual waste products were primarily used to identify specific compounds. The estimated values, however, did not consider diet and hygiene, which could significantly alter composition.

Allowable contaminant levels

Even under the best conditions, gaseous contaminants can be released in substantial amounts from human, material, and process sources within a closed system. Criteria are required that will permit establishment of safe concentrations for the contaminant mixtures formed. The concentrations specified would necessarily depend on the nature and length of exposure and the contaminants present.

Allowable concentrations, referred to as *Threshold Limit Values* (TLV's) (1964), have been established for periodic industrial-type exposures to single contaminants. Criteria have not been established, however, that permit determination of the toxicological hazard posed by continuous exposures to contaminant mixtures.

Since complete toxicological evaluations have not been possible, contaminants have been "engineered out" through source control and utilization of oversized removal equipment. This approach has been successful in manned systems such as nuclear submarines, space vehicles, and space simulators where exposure times were short or power requirements were not critical. The approach is unsatisfactory for long-term confinements in space vehicles or other systems where minimum power usage is essential. Optimum design of these systems requires the establishment of allowable contaminant concentrations.

Toxicological data

The necessity for toxicological criteria was first recognized in connection with industrial exposures. As a result, toxicological data have been collected on most industrially important compounds through animal experimentation and human exposure experience (Patty, 1963). Threshold limit values have been derived from the data. These values are time-weighted atmospheric concentrations and represent levels at which it is believed that most workers can be repeatedly exposed, 8 hrs. per day, 40 hrs. per week, without adverse effect.

Attendant with continuous human confinements in closed systems has been a need to establish safe concentrations for the contaminant mixtures present. The Navy has been confronted with the problem, particularly since the advent of the nuclear submarine. Animal experiments have been undertaken to determine a 90-day continuous exposure level for individual contaminants found in submarine atmospheres (Siegel, 1965). The data, however, do not permit assessment of the true toxic hazard posed by contaminant mixtures.

The Air Force and the National Aeronautics and Space Administration are interested in obtaining allowable concentrations applicable to space confinements. Rats, mice, and monkeys have been exposed for 90 days to single contaminants and mixtures (Sandage 1961a,b). The experiments, conducted at concentrations corresponding to the TLV's, demonstrated that these values are not applicable as continuous exposure levels. Recently, an altitude test facility designed to permit toxicological experimentation under simulated spacecraft conditions was completed at Wright-Patterson Air Force Base (Thomas, 1965). Initial experiments were conducted to determine the toxicity of carbon tetrachloride, ozone, and nitrogen dioxide in a 5-p.s.i.a. pure-oxygen environment (McNerney, 1965). The test data permit establishment of 2-week exposure levels for these compounds.

Results from human exposures in space simulators, space-craft, and submarines have provided a good basis for evaluating the potential contaminant hazard associated with long-term confinements. The Manned Environmental System Assessment (MESA) program conducted by The Boeing Company for the National Aeronautics and Space Administration demonstrated the seriousness of the trace contaminant problem (Page *et al.,*1964). The initial simulator test, MESA I, was aborted after 5 days due to equipment failure and to nausea of some crew members. Strong odors were present in the chamber which possibly contributed to the sickness. The contaminants listed in Table 1 were identified in desorbate from charcoal and silica gel exposed to the chamber atmosphere (Page *et al.,* 1964; Saunders, 1964). The system was redesigned with careful materials selection to minimize odorous and other gaseous contaminants. The contaminant control system, particularly, was redesigned with ample capacity. The second test, MESA II, was completely successful in maintaining 5 men for 30 days in a closed,

TABLE 1

CONTAMINANTS IDENTIFIED IN CHARCOAL
AND SILICA GEL DESORBATES[a]

Charcoal desorbate	
Acetaldehyde	Ethyl chloride
n-Butane	Freon-12
iso-Butylene	*n*-Pentane
Carbon dioxide	*iso*-Pentane
Carbon disulfide	Phosgene
Carbonyl sulfide	Propylene
Chloroacetylene	Trichloroethylene
Dichloroacetylene	Vinylidene chloride
Ethanol	

Silica gel desorbate	
Acetone	Methanol
Ammonia	2-Methyl butene
iso-Butylene[b]	Methyl chloride
Ethanol	Methyl ethyl ketone
Ethyl chloride[b]	Trichloroethylene[b]
Ethyl ether	

[a] The charcoal and silica gel were exposed to the chamber atmosphere in the MESA I project.

[b] Contaminants common to charcoal desorbate.

TABLE 2

CONTAMINANTS IDENTIFIED IN DESORBATE
FROM CHARCOAL: MESA II [a]

Acetaldehyde	Methanol
Butane	Methyl formate
Carbon dioxide	Nitric oxide
Ethanol	Nitrogen dioxide
Ethyl nitrite	Water

[a] The charcoal was exposed to the chamber atmosphere during the first day of the test.

self-contained environment. Contaminants identified (Saunders, 1964) in the desorbate from charcoal exposed during the first day of the second test are listed in Table 2. Comparison between Tables 1 and 2 shows the effectiveness of contaminant management in reducing the number of atmospheric contaminants.

Man's ability to maintain himself within an artificial environment has also been demonstrated in operational systems. Mercury and Gemini flights have shown that it is possible to maintain a habitable environment in space. Even though 60 different contaminants were recovered from the atmospheres of Project Mercury spacecraft (Saunders, 1963), toxicological problems did not develop. Since time-concentration history was not obtained, however, establishment of allowable limits from the data was not possible.

The success of nuclear submarines has probably been the most important single factor showing the feasibility of long-term confinements in closed systems. Numerous contaminants have been detected in submarine atmospheres (Carhart and Piatt, 1963). Others have undoubtedly been present, but were not detected. The contaminants, however, have not presented a significant hazard and, consequently, have been maintained at levels below what would be termed allowable concentrations. Complete qualitative and quantitative analyses of the atmospheres would have permitted establishment of continuous exposure levels. Although this was not done, much can be inferred simply from the success achieved.

The contaminants found in submarine atmospheres should exhibit toxicological action typical of contaminants present in any closed system. Consequently, a habitable closed system should result if it is designed by direct scaling from the submarine system. Space vehicle design would undoubtedly be far from optimum if approached in this manner. A "standard" design would be provided, however, which could be refined from information developed in contaminant management programs.

Estimated allowable concentrations

The lack of established allowable concentrations for potential spacecraft contaminants has necessitated the estimation of the values from available data. The literature was reviewed and allowable continuous exposure levels were proposed for individual contaminants on the basis of biochemical, physiological, toxicological, and industrial hygiene data. Safe mixture concentrations were estimated by applying a method suggested by Elkins (1962) for periodic industrial exposures. The contaminants were placed into one of three groups:(1) narcotics and asphyxiants; (2) respiratory irritants; or (3) systemic poisons. Chemicals in each group were assumed to exhibit additive toxicological action such that a safe mixture would exist when the ratios of actual to allowable concentration totaled less than unity.

The Hine Laboratories, San Francisco, were employed to conduct an independent evaluation. A comprehensive study was made which provided estimated continuous exposure levels for 175 probable space vehicle contaminants (Hine and Weir, 1965). Safe mixture limits were determined by assuming additive action within each of 13 different pharmacological groups. An example of the results from this study is presented in Table 3. The industrial threshold limit values are presented for comparison.

Required experimentation

Estimated allowable concentrations were obtained only for use as a guide to design. Experimentation is required to verify the estimates, particularly for contaminant mixtures where such phenomena as synergism, antagonism, potentiation, or sensitization could be important.

The Hine Laboratories proposed an experimental program designed to provide the required data. The program included studies to determine chronic toxicity of individual contaminants, toxicity of mixtures formed from contaminants with the same pharacological activity, possible interactions between contaminants with different pharmacological activity, uptake and storage, multistress effects, and odor effects. The program would involve animal experimentation and manned tests in simulated space chambers. Long-term experimental programs conducted by the Air Force, NASA, Navy, and U. S. Public Health Service are similar in scope and will

TABLE 3

ESTIMATED CONTINUOUS EXPOSURE LEVELS
FOR POSSIBLE SPACE VEHICLE CONTAMINANTS

Compound	8-Hr. TLV (p.p.m.)	Continuous exposure level (p.p.m.)	Autonomic N.S.	Blood	Cardiovascular	CNS depressant	CNS stimulant	Enzyme inhibitor	Hemopoetic tissue	Hepato agent	Mucous membrane	Nephro agent	Peripheral N.S.	Respiratory	Simple asphyxiant
Acetone	1000	500				×					×			×	
Ammonia	50	25					×				×			×	
Benzene	25	5			×	×			×		×			×	
Carbon disulfide		2					×	×					×		
Carbon monoxide	100	25	×												
Methane	1000	2500				×									×
Phenol	5	1	×	×						×		×			
Toluene	200	50				×				×	×	×			
Trichloroethylene	100	20								×	×	×		×	
m-Xylene	200	50								×	×			×	

provide required data. However, program costs, the large amounts of data, and the speed with which data can be obtained will delay the final establishment of toxicological criteria for quite some time.

In the interim, experiments have been proposed that can furnish preliminary toxicological evaluations of space vehicle atmospheres. Dr. H. E. Stokinger has suggested a manned system test where prototype subsystems and materials in actual amounts anticipated for use would be placed in a simulated space chamber (Stokinger, 1965). If a safe environment were indicated from unmanned tests and animal tests, a manned test would be conducted under close medical supervision and with adequate monitoring devices. The experiment would establish safe contaminant levels and control system operating characteristics applicable to the actual space vehicle. This approach would furnish a direct evaluation of an environment before system construction. Final evaluations would be made on the basis of the information obtained and available toxicological criteria.

Control system requirements

Trace atmospheric contaminants must be controlled to concentrations below the allowable levels. Cabin leakage and purge are effective control techniques but can carry large weight penalties, particularly for long missions. Control system weight can be reduced by using air processing equipment for contaminant removal or for conversion to less hazardous compounds. Systems considered for this application include activated charcoal adsorbers, chemisorbents, and catalytic oxidizers.

Activated charcoal adsorbers are capable of removing most trace atmospheric contaminants and thus serve as a basic system. Chemisorbents and a catalytic oxidizer can be used to control specific gases not readily adsorbed on charcoal. For example, ammonia is removed by chemisorbents, and gases such as carbon monoxide, hydrogen, and methane are converted to nonhazardous compounds by a catalytic oxidizer.

Design parameters for the various control systems can be established from knowledge of contaminants, generation rates, and allowable contaminant concentrations. Process air flow rate, L, for steady-state operation can be calculated by simultaneous solution of the following set of $n + 1$ equations:

$$c_i = G_i/L, \quad i = 1, \cdots, n \tag{1}$$

$$\sum_{i=1}^{n} \frac{c_i}{T_i} = 1.0 \tag{2}$$

where G is the generation rate, T the continuous exposure level for individual contaminants, and c the allowable contaminant concentration in the mixture. Subscripts refer to individual contaminants, n being the total number within a pharmacological group. The calculation is made for each pharmacological group of contaminants and the maximum process air flow rate obtained is the design value. If cabin leakage is present, this is subtracted from L in order to obtain the flow through removal equipment. When two or more control systems require different air flow rates, values are selected for L in all but one system. Process air flow rates are calculated for that system by following the foregoing procedure. By repeating the calculation, a trade study can be conducted to optimize the total contaminant control system.

The calculations assume complete removal of contaminants by control systems. Where this assumption does not apply, removal efficiencies must be included in Eq. 1. Also, the equations must be expanded when new contaminants are formed by reactions within removal equipment. For example, results from the MESA I test indicated that dichloroacetylene was formed by the reaction of trichloroethylene with sodium superoxide (Saunders, 1965). If similar conversions are possible in the space system under development, equations would be included that describe the generation of these contaminants.

Sizing of adsorption equipment requires knowledge of capacity and adsorption rates in addition to process air flow rates. This information is available for many single contaminants but has not been developed for adsorption from low concentration contaminant mixtures. Design parameters have been experimentally determined for catalytic oxidizers. Catalysts should be developed, however, that will permit operation at lower temperatures and possibly reduce contaminant formation.

SUMMARY

A program has been developed for the management of potential atmospheric contaminants in manned space vehicles and other closed environmental systems. The success of the program in its application depends on securing reliable data in several areas of work. Contaminants must be identified and their production rates from materials and man determined. The total amount of each material used must be determined. The tolerance limits for continuous exposure of humans to complex mixtures of toxic compounds must be established. This information will permit the design and construction of efficient air purification equipment.

Experimental programs are in progress in many governmental, university, and industrial laboratories that will yield these data. The Boeing Company has experimental programs in progress in each area. Close cooperation and understanding are essential among people in the several disciplines involved—toxicology, chemistry, engineering, and design—in order to accomplish the desired results.

Acknowledgment

The authors wish to express their appreciation to Mr. M. T. Braun for his direction of the contaminant management program.

The assistance of numerous individuals in the Bioastronautics, Flight Technology, and Structures and Materials Technology Organizations is also gratefully acknowledged.

REFERENCES

Carhart, H. W., and V. R. Piatt, eds. (1963). The present status of chemical research in atmosphere purification and control on nuclear-powered submarines. NRL Rep. 6053, U.S. Naval Research Lab., Washington, D. C. (December).
Elkins, H. E. (1962). Maximum allowable concentrations of mixtures. American Industrial Hygiene Assoc. J., 23 (2): 132.
Hine, C. H., and F. W. Weir (1965). Probable contaminants and their recommended air levels in space vehicles. Boeing Doc. D2-90731-1, The Boeing Company, Seattle, Washington (February).
McNerney, J. M. (1965). Preliminary results of toxicity studies in 5 p.s.i.a.-100% oxygen environment. Paper presented at the Conf. on Atmospheric Contamination in Confined Spaces, Dayton, Ohio (March 31, April 1-2).
Page, R. N., C. Dagley, and S. Smith (1964). Manned environmental system assessment (MESA) program—Final Report. Boeing Doc. D2-90487-5, The Boeing Company, Seattle, Washington (June).
Patty, F. A., ed. (1963). Industrial hygiene and toxicology, 2nd rev. ed., Vol. II. Interscience, New York.
Sandage, C. (1961a). Tolerance criteria for continuous inhalation exposure to toxic material (I). Effects on animals of 90-day exposure to phenol, CCl_4, and a mixture of indole, skatole, H_2S, and methyl mercaptan. ASD-TR-61-519(I). Wright-Patterson Air Force Base, Ohio (October).
_____(1961b). Tolerance criteria for continuous inhalation exposure to toxic material (II). Effects on animals of 90-day exposure to H_2S, methyl mercaptan, indole, skatole. ASD-TR-61-519(II), Wright-Patterson Air Force Base, Ohio (December).
Saunders, R. A. (1963). Analysis of atmospheric contaminants recovered from MA-9 (Faith VII). Naval Research Lab. Letter Rep. 6110-298A:RAS:bs, Washington D. C. (October 22).
_____(1964a). Atmospheric contamination in the aborted MESA environmental test chamber study. Naval Research Lab. Letter Rep. 6110-2:RAS:bs, Washington D. C. (January 8).
_____(1964b). Analysis of project MESA test chamber atmospheres. Naval Research Lab. Letter Rep. 6110-96A:RAS:bs, Washington D. C. (May 8).

_____(1965). The source and identity of the toxicant in the project MESA atmosphere. Naval Research Letter Rept. 6110-247A:RAS: vmg, Washington, D. C. (Oct. 8).

Siegel, J. (1965). Review of ambient pressure animal exposure data from selected navy compounds. Paper presented at the conf. on Atmospheric Contamination in Confined Spaces, Dayton, Ohio (March 31, April 1 and 2).

Gorban, C. M., I. I. Kondrat'yeva, and L. T. Poddubnaya (1964). Gaseous activity products excreted by man when in an air-tight chamber. *In* Problems of space biology, (N. M. Sisakyan and V. I. Yazdovskiy, eds.). Joint Publications Research Service, Washington, D. C. (June).

Stokinger, H. E. (1965). Personal communication (February).

Thomas, A. A. (1965). Chamber equipment design considerations for altitude exposures. Paper presented at the Conf. on Atmospheric Contamination in Confined Spaces, Dayton, Ohio (March 31, April 1-2).

Threshold Limit Values (1964). Adopted at the 26th Ann. Meeting Am. Conf. Governmental Industrial Hygienists. Philadelphia (April 25-28).

Webb, P., ed. (1964). Bioastronautics data book. NASA SP3006.

9.

ALGAL BIOREGENERATIVE SYSTEMS

R. L. MILLER · C. H. WARD

Environmental Systems Branch
USAF School of Aerospace Medicine
Brooks Air Force Base, Texas

INTRODUCTION

Algae may be used for partial regeneration of man's requirements for life in a closed environment. Feasibility has been demonstrated with model systems, but established principles of algal metabolism impose severe restrictions on the design of thermodynamically efficient, low-volume and low-weight algal gas exchangers. Review of available data on photosynthetic gas exchangers now permits verification of design parameters predicted almost 10 years ago. Experimentally achievable values of electrical efficiency are only a fraction of the theoretical. Significant improvement over or attainment of theoretical values will require major improvement in the conversion of electrical energy into light energy or conversion of light energy into chemical energy by the green plant. Development of a basic design theory would be greatly simplified by a definite mission-oriented goal, for example, a planetary base. At present there is no material advantage among existing algal exchangers since criteria used for design require compromises of weight, volume, and power. Most algal gas exchangers are inadequately described. More experimental data, obtained by extensive operation of prototype systems, are needed for accurate logistic evaluation. Weight, volume, and power flexibility may ultimately be of advantage in the design of life-support systems for specific space missions, provided long-term reliability can be demonstrated. This paper attempts to compare and evaluate the available information on algal photosynthetic gas exchange systems.

Pursuit of our national goals in space exploration will eventually require man's long-duration tenancy of celestial vehicles and planetary bases. Requirements for life support could be met through expenditure of stored supplies and by regeneration and reuse of the waste products of human metabolism. The logistic necessity of regeneration for extended space missions is well documented (Welch, 1961); however, the mission time at which regeneration will be logistically profitable is subject mostly to debate.

Several regenerative processes, chemical and biological, have been suggested and investigated for CO_2 removal, O_2 generation, waste control, and nutritional support in closed environments (cf. Acker and Stern, 1960; Anonymous, 1960; Bongers, 1964; Golueke, 1962; Ingram, 1958; Rousseau, 1964). Some have unique characteristics for application to one or more of the problems involved.

The biological approach has as its pattern the complex multi-organismal life-support system that has evolved on earth. This system, stable because of its size and the contributions of many organisms, is powered by light energy from the sun and based on green plant photosynthesis. This approach to space life support has been dated by Myers (1962) as originating in 1951 when Dr. Heinz Specht suggested that human respiratory requirements might be managed by photosynthesis of plants (Specht, 1952). However, in developing the philosophy or rationale governing the use of green plants, particularly algae, for photosynthetic exchange and regeneration, one can do little more than paraphrase Dr. Jack Myers, for he has been our spokesman and critic (Myers, 1954, 1956a, b, 1958, 1960a, b; 1963, 1964a, b; Myers and Brown, 1961). In addition, numerous general articles have appeared on the philosophy, theory, engineering, and application of bioregenerative (mostly algal) systems (Bassham, 1954; Bates, 1961; Benoit, 1964; Beyers, 1963; Bongers, 1964; Bongers and Kok, 1964; Bowman, 1953; Burk et al., 1962; Clamann, 1961, 1963; Dyer, 1963; Enebo, 1960; Gafford, 1962, 1963; Gafford and Richardson, 1960; Gaume, 1957; Golueke, 1962; Golueke et al., 1959; Hobby, 1959; Johnson and Finn, 1963; Konikoff, 1960, 1961; Kratz, 1959; Leonard, 1960; Mattoni, 1963; Phillips, 1961, 1962b; Popma, 1962; Tischer and Tischer, 1963; Wilks, 1963). The literature on bioregeneration has been compiled (Anonymous, 1965a, b; Garrick, 1963; Gini, 1960; Ingram, 1958; Spiegler, 1963) and recently reviewed (Golueke and Oswald, 1964).

Use or misuse of ecological terminology in describing bioregenerative research has led to a general misconception in the scientific community that success of the bioregenerative approach depends on development of a biologically and chemically closed ecology with complete material balance. While this may be the ultimate goal, few consider it possible if indeed necessary. The study of complex natural biotic communities as a basis for development of materially balanced space ecologies may have merit (Odum, 1963). However, the most popular (and hopefully the most productive) approach involves selection and integration of specific organisms (processes) or groups of organisms particularly adapted to one or more phases of regeneration. This step-by-step building process falls into the area of biological engineering (Myers, 1964b). Hence, the use of algae for bioregeneration is a biological engineering effort.

The purpose of this paper is to discuss: (1) requirements for regeneration; (2) factors affecting algal growth; (3) engineering design and performance of algal exchangers; and (4) future development.

REQUIREMENTS FOR REGENERATION

Human turnover

A brief review of human turnover is requisite to a discussion of space life support (Table 1). While human turnover varies depending on diet and activity, Clamann's data (1959) are generally accepted as conservative estimates. Experimentally determined values for a relatively inactive man in a closed environment are given for comparison (Welch, 1961). The human respiratory quotient ($RQ = CO_2/O_2$) is variable between about 0.7 to 1.0. For purposes of discussion, we will accept a value of 600 liters (860 gm.) man-day as reasonable for oxygen requirement.

Table 1 shows that four partial functions are required for complete regeneration. In order of magnitude, but not necessarily of difficulty, these are: (1) recycling of water; (2) exchange of

TABLE 1

HUMAN TURNOVER IN GRAMS PER DAY

Moderately active[a]				Relatively inactive[b]			
Input		Output		Input		Output	
H_2O	2,200	H_2O	2,540	H_2O	2,268	H_2O	785
O_2	860	CO_2	980	O_2	544	CO_2	644
Food (dry)	520	Solid waste (dry)	60	Food (dry)	544	Solid waste (dry)	—[c]

[a] Data from Clamann (1959).   [b] Data from Welch (1961).
[c] Not measured.

carbon dioxide and oxygen; (3) production of food; and (4) recovery of wastes. Myers (1964b) states that 85% of man's requirements would be satisfied by accomplishing the first two partial functions. Therefore, processes for partial regeneration may ultimately be of advantage when compared to expendable systems.

Algal turnover

Algal cells (green plants) use light energy to convert carbon dioxide and water into oxygen and organic compounds required for the formation of new cell material. The first of the syntheses leading to the production of new cell material is photosynthesis. Essential features of green plant photosynthesis are light-activated photolysis of water, liberation of oxygen, and production of "assimilatory power" in the form of adenosine triphosphate (ATP) and reduced pyridine nucleotide ($TPNH_2$). ATP and $TPNH_2$ are subsequently used in nonphotosynthetic assimilation (reduction) of carbon dioxide into carbohydrates, lipids, and proteins. Carbon dioxide assimilation is now known to occur in the metabolism of most organisms; however, use of light energy for the production of assimilatory power is unique in photoautotrophic metabolism. Current concepts of photosynthesis and carbon dioxide assimilation have been expertly reviewed by Arnon (1961, 1962).

Algal turnover characteristics have been considered in detail (Myers and Brown, 1961). Myers (1960b, 1964b) has shown that elementary analysis of algal cells permits estimation of over-all metabolism and derivation of equations for cell synthesis. With urea as the nitrogen source, equivalents calculated for 100% recovery of carbon show that for each gram (dry) algae produced, 0.82 liter of carbon dioxide is assimilated and 1.0 liter of oxygen is liberated, resulting in an assimilatory quotient ($AQ=CO_2/O_2$) of 0.82. A human oxygen requirement of 600 liters/day would require the uptake of about 480 liters/day of carbon dioxide and the production of about 600 gm. (dry) of algae. Algal AQ can be varied between about 0.7 to 0.9, depending on the source of nitrogen used for cell synthesis. The need for algal AQ-human RQ balance is critical since it can be shown that a mismatch of 1% leads to accumulation or loss of 1% of the human oxygen requirement per day.

FACTORS AFFECTING ALGAL GROWTH

The physiology, biochemistry, and culture of algae have been extensively reviewed (Krauss, 1958; Myers, 1951; Tamiya, 1957) and several volumes have been devoted to discussion of these subjects (Anonymous, 1956; Brunel *et al.*, 1950; Burlew, 1953; Carpenter, 1955; Fogg, 1953; Jackson, 1964; Kachroo, 1960; Lewin, 1962; Robinette, 1962). We intend only to briefly discuss variables believed especially significant in the mass culture of algae for bioregeneration. Similar information has been presented in more elegant form (Myers, 1964b).

Species

No known algal species or strain has characteristics ideally suited to mass culture and gas exchange. Several species of *Chlorella*, *Anacystis*, *Synechocystis*, *Scenedesmus*, *Synechococcus*, and others have been evaluated (Benoit *et al.*, 1960; Dyer and Gafford, 1962; Felfoldy, 1964; Gafford and Craft, 1959). Recently, members of the Polyblepharidaceae (marine flagellates) were evaluated for productivity in mass culture (Eppley *et al.*, 1964a,b). Of the unicellular forms, species of *Chlorella* have been used most frequently. Their selection was undoubtedly based on ease of culture and the extensive literature on their general physiology. Most strains of *Chlorella* tolerate wide variations in their environment and are easily adapted to mechanized mass culture. Sorokin's (Sorokin and Myers, 1953) thermotolerant (39°C.) *Chlorella pyrenoidosa* Chick TX71105 is now used almost to the exclusion of other potentially useful forms. Many investigators consider TX71105 a rugged, dirty beast, but the best choice pending establishment of more definitive selection criteria.

Carbon dioxide

Maximum photosynthesis and growth of Chlorellas can be maintained at carbon dioxide liquid phase tensions from about 1 to 40 mm. (Davis *et al.*, 1953; Myers, 1964b; Myers and Brown, 1961; Nielsen, 1952). The upper and lower limits are not well defined, but aeration of algal cultures with 5% carbon dioxide in air is fairly routine laboratory practice. Lower concentrations are often desirable for control of pH (Galloway and Kraus, 1961) and may be used effectively depending on the efficiency of liquid-gas transfer. The lower levels of carbon dioxide tolerable to man ($\approx 1.0\%$) present problems in gas exchanger design since large liquid-gas contacting surfaces are required for satisfactory phase transfer. An alternative is the use of concentrated carbon dioxide at a rate just sufficient to meet the carbon demand (Tew *et al.*, 1962). However, in tightly closed systems, oxygen toxicity may then become a problem.

Oxygen

Oxygen toxicity, commonly referred to as the Warburg effect, has been extensively investigated but not adequately explained. Rate of photosynthesis decreases in cultures in equilibrium with oxygen pressures greater than 160 mm. (Turner and Brittain, 1962). Myers (1964b) states that casual observations in his laboratory suggest that oxygen pressures above 700 mm. lower the growth rate of *Chlorella pyrenoidosa*. Experiments (unpublished) performed in our laboratory in 1962, and subsequently confirmed by Dyer *et al.*, 1963, demonstrated that oxygen pressures above 700 mm. can decrease the growth of TX71105 up to 75%. Oxygen toxicity should not be of practical concern in the development of algal exchangers, provided adequate liquid-gas transfer can be accomplished at low carbon dioxide pressures.

Nutrients

There is an extensive literature on the mineral nutrition of algae (Burlew, 1953; Eyster, 1964; Krauss, 1958). Green algae require, in addition to carbon dioxide and water, a source of fixed nitrogen (urea, ammonia, or nitrate) and mineral salts. Most media contain potassium, magnesium, sulfate, and phosphate ions in macro quantities, plus 1 to 30 p.p.m. of some 8 to 10 microelements. A host of nutrient formulations has been published (Lancaster and Tischer, 1962), but there is no best recipe. Media for green algae generally differ only in the salt form and concentration of the mineral elements. In algal mass cultures, the requirement for each element will depend on the production rate; however, most commonly used media contain luxuriant quantities of all elements except nitrogen (Gafford and Craft, 1959; Gaucher *et al.*, 1960b; Krauss and Thomas, 1954; Myers, 1957).

Assuming an average of 8% nitrogen and 5% ash on a dry weight basis, production of 600 gm. algae/day will require about 100 gm. of urea (48 gm. N) and about 30 gm. of mineral salts. Reclamation of human wastes (average diet) would provide about 26% of the nitrogen, 15% of the magnesium, 27% of the phosphorus, 10% of the iron, and all of the microelement requirement (Golueke and Oswald, 1964). Complete recycling of daily wastes from a human on a total algal diet would provide the complete mineral requirement for the growth of 600 gm. algae/day (Myers, 1964b). Human wastes, untreated (Lynch *et al.*, 1964; Moyer, 1962), sewage plant processed (cf. Golueke and Oswald, 1964), and electrolyzed (Brown *et al.*, 1964) have been used to supply all or part of the nutrient requirement of algal cultures.

Practical management of continuous algal mass cultures for bioregeneration will eventually require nutrient reconstitution and reuse of effluent medium following algal harvest. The early work by Krauss and Thomas in 1954 formed the framework for future development in this area. Reconstitution methodology has not since been significantly improved (Dyer and Wildman, 1964; Leone, 1963) and is still under active investigation (Eyster, 1965).

Temperature

Each alga has a fairly wide temperature range in which growth can occur (Krauss and Osretkar, 1961). However, optimal growth usually occurs within a much narrower range (2 to 3°C.). Most common mesophilic strains have optima at about 25°C., while Sorokin's thermotolerant *Chlorella* grows best at 39°C. Characteristics of a thermophilic blue-green alga with a temperature optimum at 52°C. have been reported (Dyer and Gafford, 1961).

At one time, high temperature algae were believed to use high intensity light more efficiently and to be more productive in mass culture (Burk *et al.*, 1962). The argument was based on the high values for light saturation and attendant high maximum doubling rates. Unfortunately, high temperature strains have not proved to be significantly more productive in mass culture (Myers and Graham, 1961). However, they do offer an advantage in that less cooling capacity is required to maintain acceptable temperature limits.

Light

The effect of light is probably the most important consideration in the design of photosynthetic exchangers. Two properties of light energy are important to algal growth and metabolism: spectral quality and intensity (irradiance or illuminance).

Spectral quality is defined by the absorption spectrum for the chlorophylls and other photosynthetically active pigments. Chlorellas typically absorb energy of wavelengths from less than 300 to over 700μ with peaks in the blue and red regions of the spectrum (Myers and French, 1960). Since prolonged exposure to ultraviolet radiation is inhibitory to growth and cell division (Kok and Bongers, 1961), the allowable spectral limits for sustained productivity fall in the range from 400 to 700μ.

The effects of light intensity on algal growth and photosynthesis have been studied extensively by several investigators (Krauss and Osretkar, 1961; Myers, 1946; Sorokin and Krauss, 1962). We will briefly consider light effects in optically thin and optically

dense cultures. Thin cultures define basic limitations at the cellular level. Dense cultures are of interest for practical application.
In optically thin cultures, the effect of irradiance on the specific
growth rate is given by a response curve such as Figure 1. At
low intensities, the rate increases with increasing intensity, and
efficiency, given by the slope of the curve, is maximal. At higher
intensities the rate becomes light saturated, and at very high intensities may be inhibited. For any algal species, the character of
the light curve is not unique but depends upon temperature (Tamiya
et al., 1953), previous history (Myers and Graham, 1959), ances-
try (Sorokin, 1958), and age or degree of cell development (Sorokin,
1960).

An important feature of the light curve is the relatively low
values of irradiance at which growth rate becomes light saturated.
Phillips and Myers (1954) measured growth rate of *Chlorella
pyrenoidosa* (25°C.) as a function of irradiance in steady-state
experiments and obtained a saturation irradiance of about 35 kilo-
erg/sq. cm.-sec. (3.5 mw./sq. cm.). Sorokin and Krauss (1962)

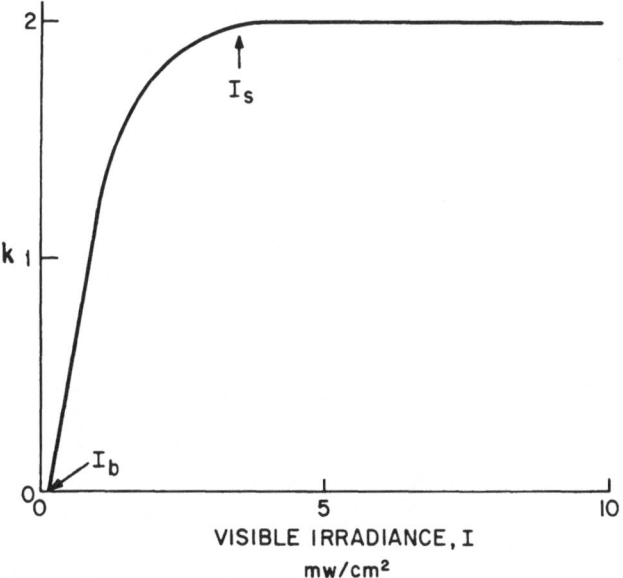

Figure 1. Specific growth rate *k* as a function of
irradiance. I_b denotes the irradiance required to balance respiration, I_s the irradiance at. which growth
rate becomes light saturated. Taken from Myers
(1964b), and based on data of Phillips and Myers (1954).

measured growth versus illuminance for five different strains of algae and found saturation intensities from 250 ft.-c. for *Chlorella vulgaris* (25°C.) to 1400 ft.-c. for a high temperature strain of *Chlorella* (39°C.). In practical application, this presents a severe limitation because most light sources (including sunlight) have intensities much greater than these values. A second and related limitation imposed by light saturation is that maximum efficiency of energy utilization by algae necessarily occurs at low values of intensity, where the rate is light limited. The maximum efficiency at which algae convert visible light energy into cellular (chemical) energy lies in the range from 18 to 22% (Kok, 1952; van Oorschot, 1955).

Maximum productivity and greatest efficiency are obtained in light-limited optically dense cultures. In optically dense cultures, light intensity decays with optical thickness by an absorbance curve such as Figure 2. Gradations in light with culture thickness give rise to gradations in cellular photosynthetic rates which are governed by a response curve such as Figure 1. Hence in dense

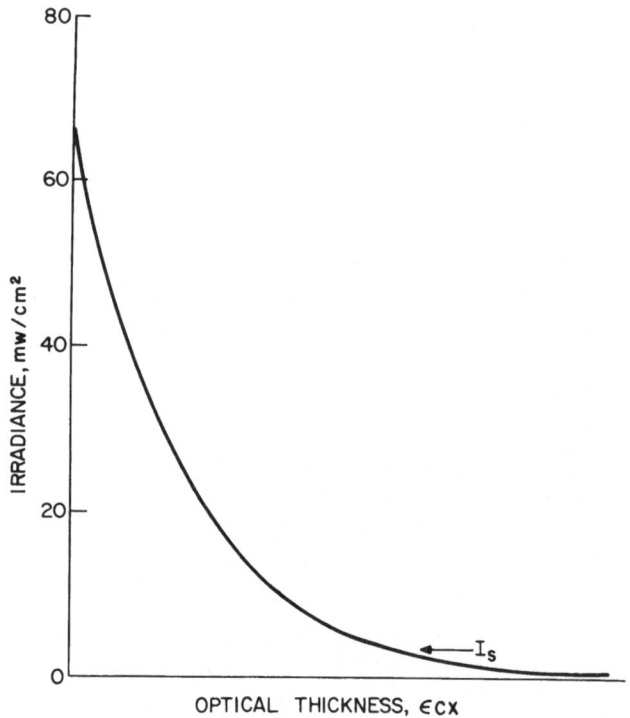

Figure 2. Irradiance decay in algal cultures as a function of optical thickness, ϵcx.

cultures, the over-all rate is a composite of the cellular rates for each differential increment of culture depth, and may be determined by integration. Several mathematical treatments have been made on this basis (Fredrickson, 1961; van Oorschot, 1955; Tamiya *et al.*, 1953), but no one model has yet accounted for all of the many factors involved. The simplest analysis results in a form of the Bush equation:

$$R = \frac{E_m I_s}{J} \left(\ell n \frac{I_0}{I_s} + 1 \right)$$

where I_0 is the incident irradiance, I_s the saturating irradiance E_m the maximum efficiency of light utilization, J the heat of combustion of algae, and R the over-all growth rate per unit area. The Bush equation demonstrates, in a qualitative sense, the over-all effect of increasing irradiance on dense culture performance. The inadequacies and uncertainties in its derivation have been given by Myers (1964b). The effect of light in dense cultures will be discussed in more detail in a later section where we have used the Bush equation to aid in the correlation and interpretation of photosynthetic gas exchanger results.

Contaminants

Algal mass cultures are maintained free of contaminants only under carefully controlled conditions and then with considerable difficulty. Many problems associated with the culture of algae have been attributed to the presence and buildup of contaminants. Detrimental effects of protozoan predators have been reported (Burlew, 1953; Tamiya, 1957). Large numbers of bacteria (10^6 to 10^9 viable cells/ml.) are frequently associated with algae in dense laboratory cultures (Krauss and Thomas, 1954; Ward *et al.*, 1963; Zuraw and Adamson, 1963). Ward *et al.* (1964) showed that contaminant bacteria grow on algal excretory products and that their growth is a function of and closely parallels algal growth. Although representing 1% or less of the total culture mass, four of six bacterial isolates tested were found to significantly reduce algal growth. It was also shown that some bacteria have competitive advantage in algae-bacteria cultures owing to selective utilization of excretory products. However, Blasco (1965) failed in attempts to grow bacteria on effluent from axenic algal cultures and postulated a pathogenic relationship to explain the growth of bacteria.

In recent experiments performed in our laboratory, most soil and air bacteria tested survived in algal cultures for several days but did not increase in numbers (Vela *et al.*, 1965). Two human enteric pathogens, *Salmonella typhi* and *Salmonella paratyphi*,

grew in algal cultures for extended periods; this finding suggests
that it may be necessary to biologically isolate man from an algal
regeneration system and sterilize bioregenerative products intended for human consumption.

A frequent qualitative observation is that foaming, sticking,
and clumping become more problematic when cultures are heavily
contaminated. Chemical control of contaminants may be possible
(Blasco, 1965; Galloway and Krauss, 1959), but not advisable if the
bacteria are metabolizing compounds that become autotoxic in high
concentrations, for example, in a closed water cycle. Additional
research is needed to further characterize the role of contaminant
bacteria in algal cultures.

Pressure, radiation, magnetic fields, acceleration,
vibration, gravity

These variables are often discussed relative to space research;
however, little is known about their influence on algal growth.

Algae are relatively insensitive to pressures ranging from the
vapor pressure of water to an upper limit that probably exceeds 1
atm. (Myers and Brown, 1961). Pressure has a direct effect on solubility of gases in water and, hence, could have an indirect effect on
algal growth (Hannan, 1964a).

Effects on ionizing and nonionizing radiation on *Chlorella* and
other algae have been reviewed (Godward, 1962; Kok and Bongers,
1961). Ultraviolet effects have been extensively studied while effects
of most ionizing radiations are less well known. Types of damage
generally differ, depending on the form, dose rate, and total dose
of the radiation received. *Chlorella* is fairly tolerant to acute exposure to high doses of most radiations (Godward, 1962), but sensitive to chronic exposure to low doses of gamma radiation (Posner
and Sparrow, 1964). Acute exposures may cause mutation or even
death of individual cells, but have little or no effect on the subsequent growth of the surviving population. Resistance to acute
exposure is of limited practical importance to space application
since more sensitive forms, especially man, must be protected in
high radiation environments. Chronic exposure effects may become
limiting in long-term maintenance of culture stability, even though
survival of less productive mutant forms is unlikely (Mitchell, 1960).

Magnetic fields are known to cause a variety of effects in several different organisms (Barnothy, 1962; Beischer, 1963b; Davis
et al., 1962) and could be important in manned space flight (Beischer,
1963a. Information on the effects of magnetic fields on algae is
limited and conflicting. Halpern and Konikoff (1964) report that
magnetic fields below 1,000 gauss inhibit growth while fields up to

20,000 gauss (highest tested) are stimulatory. Hannan (1964b) however, found no effect of 10,000 gauss on photosynthesis or 100,000 gauss on respiration of TX71105. This apparent conflict prohibits conclusions at this time.

Experiments performed in ballistic missiles and earth satelites provide the only information on effects of acceleration, vibration, and gravity. Phillips (1962a) was unable to detect changes in viability, genetic stability, or growth rate of *Chlorella* cultures, compared to ground controls, following a 50-hr. flight in a Discoverer satellite. Phillips' results have been confirmed in similar experiments performed by Russian workers (Semenenko and Vladimirova, 1963) and by Ward and Guerra (1962). Gafford and Bailey (1963) found no change in photosynthesis of a *Chlorella* culture during launch and a brief period of subgravity flight. Additional information on these variables should be forthcoming from the NASA and Air Force biosatellite programs (Stambler, 1965).

ENGINEERING DESIGN AND PERFORMANCE

The design of an algal gas exchange system must accommodate the biological requirements outlined in the previous section. Within this framework, several engineering objectives are possible. For support of the human in space, reliability must be paramount. However, minimum weight and volume, as well as minimum power requirement, are logistically important. Numerous algal systems have been designed to accomplish some but not all of these objectives. As a result, there is a wide diversity of algal exchanger apparatus. To provide guidelines for future development, it is instructive to review these devices and attempt to arrive at the current status of development. Accordingly, Table 2 lists the important design parameters and performance data for a number of different algal culture systems. Our criteria for selection were adequate description, continuous culture capability, and potential photosynthetic exchanger application.

Light sources

Fluorescent and incandescent (primarily Quartzline) lamps have been used in exchanger designs. Both of these lamps provide light energy of adequate quality for photosynthesis, but differ appreciably in the engineering requirements imposed by their use.

Fluorescent lamps are bulky but have the desirable characteristics of long life (10,000 hr.), relatively low surface illuminance

TABLE 2, PART 1

ALGAL PHOTOSYNTHETIC GAS EXCHANGERS

Comparison of Design Parameters, Yields, Efficiencies, and One-Man Support Requirements[a]

References	Algal species[b]	Temp. (°C.)	Culture chamber design	Total vol.[c] (liters)	Mode oper.[d]	Illumination system			Incident illuminance[g] (ft.-c.)
						Lamps		Position[f]	
						No.	Type[e]		
Bovee et al., 1962	C.p.	39	Panel	380	SC	369	30w-FL	E	1,000
Bowman & Thomae, 1960	C.p.	38	Annulus	4	SC	5	40w-FL	I & E	350
Casey et al., 1963	C.p.	39	Rect. tank	4560	SC	124	215w-FL	I	—
Cook, 1951	C.p.	25	Cylinder	10	C	3	100w-FL	E	—
Davis, 1953	C.p.	25	Annulus	0.1	SC	6	300w-IL	E	7,000
Eley & Myers, 1964	C.e.	25	Annulus	0.36	C	9	15/20w-FL	I & E	—
Gafford & Fulton, 1962	C.p.	40	Dome	9	SC	—	Sunlight	E	Var.
Gafford & Fulton, 1962	C.p.	40	Panel	14	SC	—	Sunlight	E	Var.
Gaucher et al., 1960a	C.p.	37	Annulus	4	SC	1	1500w-QL	I	10,000
Gaucher et al., 1960a	C.p.	39	Annulus	11	C	3	1500w-QL	I	35,000
Hamman et al., 1961	C.p.	38	Rect. tank	22.8	C	3	1500w-QL	I	15,000
Hamman et al., 1963	C.p.	38	Cylinder	6.2	C	6	1500w-QL	I	23,000
Hemerick, 1963	C.p.	39	Lenticular	12	B	1	1500w-QL	I	11,000
Hemerick & Benoit, 1962	C.p.	39	Annulus	12	C	1	1500w-QL	I	35,000
Krall & Kok, 1960	A.n. & Sc.	32-38	Rect. tank	182	SC	276	42w-FL	I	2,000
Leone, 1961	C.p.	38	Annulus	5	C	1	1500w-QL	I	6,000
Matthern & Koch, 1963	C.p.	38	Cylinder	2.8	C	4	1500w-QL	I	6,000
Meleshko, 1964	C.sp.	40	—	65	B	—	IL	E	9,300

References	Algal species[b]	Temp. (°C.)	Culture chamber design	Total vol.[c] (liters)	Mode oper.[d]	Lamps No.	Lamps Type[e]	Position[f]	Incident illumination[g] (ft.-c.)
Myers & Graham, 1959	C.e.	25	Cylinder	1.01	C	—	IL	E	—
Myers & Graham, 1961	C.e.	25	Cyl. (cone)	0.86	C	—	IL	E	—
Newland & Price, 1963	C.sp.	29-34	Annulus	3.0	C	2	20w-FL	I	900
Shuler, 1963	C.p.	39	Annulus	1.8	C	1	1500w-QL	I	9,000
Tew et al., 1962	C.p.	38	Annulus	1.2	C	1	150w-FL	I	—
Thacker & Babcock, 1959	C.p.	25	Glass tube	68	B-SC	120	40w-FL	E	2,000
Wallman et al., 1962	C.p.	38	Rect. tank	600	C	36	215w-FL	I	2,500
Ward et al., 1963	C.sp.	38	Panel	8	C	8	110w-FL	E	3,500
Zuraw et al., 1960	C.p.	39	Rect. tank	57	C	5	1500w-QL	I	—
Zuraw et al., 1961	C.p.	39	Ann.-dome tank	66	C	5	1500w-QL	I	—
Zuraw & Adamson, 1963	C.p.	39	Ann.-dome tank	53	C	10	1500w-QL	I	15,000

TABLE 2, PART 2[a]

Reference	S.A.[h] (sq.m.)	S.A.:Vol. Ratio[i] (sq.cm./liter)	L.P.:S.A.[j] ratio (kw./sq.m.)	Yield[k] (gm./sq.m.·day)	Yield[k] (gm./l.·day)	Efficiency[l] (gm./kw.-hr.)	Efficiency[l] (%)	One-man support[m] S.A. (sq.m.)	One-man support[m] Vol. (liters)	One-man support[m] L.P. (kw.)
Bovee et al., 1962	22.3	590	0.41	17	1.2	1.5	0.90	35	500	17
Bowman & Thomae, 1960	—	—	—	—	0.7	0.6	0.38	—	857	42

Reference	S.A.[h] (sq.m.)	S.A.:Vol. Ratio[i] (sq.cm./liter)	I.P.:S.A.[j] ratio (kw./sq.m.)	Yield[k] (gm./sq.m.·day)	Yield[k] (gm./l.·day)	Efficiency[l] (gm./kw.-hr.)	Efficiency[l] (%)	One-man support[m] S.A. (sq.m.)	One-man support[m] Vol. (liters)	One-man support[m] I.P. (kw.)
Casey et al., 1963	33.4	70	0.80	30	0.2	1.6	1.02	20	3000	17
Cook, 1951	0.44	440	0.68	11	0.5	0.7	0.45	55	1200	36
Davis, 1953	0.015	1510	119	86	13	0.03	0.02	7	46	—
Eley & Myers, 1964	0.13	3640	1.3	16.5	6.0	0.5	0.32	36	100	50
Gafford & Fulton, 1962	0.5	—	—	42	1.5	—	—	14	400	—
Gafford & Fulton, 1962	0.5	500	—	50	1.8	—	—	12	340	—
Gaucher et al., 1960a	0.13	330	11.4	202	6.7	0.74	0.47	3.0	90	34
Gaucher et al., 1960a	0.17	150	15.4	157	2.4	0.30	0.19	3.8	250	83
Hannan et al., 1961	0.23	100	19.6	120	1.2	0.25	0.16	5.0	500	100
Hannan et al., 1963	0.24	390	28.3	350	14.1	0.52	0.33	1.7	43	48
Hemerick, 1963	0.56	460	2.7	100	4.5	1.5	0.96	6.0	133	17
Hemerick & Benoit, 1962	0.11	140	13.6	180	1.4	0.79	0.50	3.3	428	32
Krall & Kok, 1960	14.6	800	0.55	34	2.7	2.6	1.66	18	222	9.6
Leone, 1961	0.14	280	8.6	115	3.2	0.56	0.36	5.2	188	45
Matthern & Koch, 1963	0.09	325	46.4	410	13.3	0.37	0.24	1.5	46	68
Meleshko, 1964	—	—	—	—	246	—	—	—	3.5	—
Myers & Graham, 1959	0.0020	20	—	71.0	0.14	—	—	85	—	—
Myers & Graham, 1961[n]	0.0017 / 0.0178	17 / 207	—	155	0.31	—	—	39	—	—
Newland & Price, 1963	0.23	1800	0.17	8.4	0.66	2.0	1.28	71	1000	12.5
Shuler, 1963	0.06	380	14.7	240	9.0	0.68	0.43	2.5	67	37
Tew et al., 1962	—	—	—	—	3.0	1.0	0.64	—	200	24
Thacker & Babcock, 1959	5.8	1530	0.82	14.4	1.4	1.2	0.77	42	429	21
Wallman et al., 1962	10.7	220	0.72	65	—	3.5	2.24	9.2	—	7

Reference	S.A.[h] (sq.m.)	S.A.:Vol. ratio[i] (sq.cm./liter)	I.P.:S.A.[j] ratio (kw./sq.m.)	Yield[k] (gm./ sq.m.·day)	(gm./ l.·day)	Efficiency[l] (gm./ kw.-hr.)	(%)	One-man support[m] S.A. (sq.m.)	Vol. (liters)	I.P. (kw.)
Ward et al., 1963	1.0	2000	0.88	50	6.0	2.4	1.53	12	100	10.4
Zuraw et al., 1960	0.30	60	24.6	165	1.1	0.28	0.18	3.6	545	89
Zuraw et al., 1962	0.95	145	7.9	100	1.2	0.54	0.35	6.0	500	46
Zuraw & Adamson, 1963	0.95	180	15.8	121	2.2	0.32	0.22	5.0	273	78

[a] No distinction is made in this table between original (cited) and calculated data.

[b] Abbreviations: C.p., *Chlorella pyrenoidosa*; C.e., *Chlorella ellipsoidea*; A.n., *Anacystis nidulans*; Sc, *Scenedesmus*; C.sp., *Chlorella* species (strain unspecified).

[c] Total volume includes both culture vessel and external recirculation volume when applicable.

[d] Mode of operation: B, batch; SC, semicontinuous (periodic harvest); C, continuous (continuous harvest).

[e] Type of illumination source including nominal wattage: FL, fluorescent lamp; QL, Quartzline lamp; IL, incandescent lamp.

[f] Position of lamp with respect to culture: E, external; I, internal.

[g] Incident illuminance upon culture vessel. All data are cited figures.

[h] Area of culture exposed to the incident irradiance.

[i] Surface area per unit illuminated culture volume.

[j] Illumination power input per unit surface area.

[k] Yields are grams dry algae per unit illuminated surface area and per unit total culture volume.

[l] Efficiency is over-all conversion of electrical energy into algal cells; percent efficiency is based on 5.5 kcal./gm. dry algae.

[m] Based on a one-man oxygen requirement of 600 liters/day (equivalent to the production of 600 gm. dry algae/day). S.A., illuminated surface area; Vol., total culture volume; I.P., input (electrical) power for illumination.

[n] This and the foregoing paper demonstrate the use of a diffusing cone for attenuation of irradiance. Surface areas are base (input) over lateral (output) of the cone. Surface-to-volume ratios are based on input and output light areas. Yield is based on input area.

(4,000 ft.-c.), and high efficiency (18 to 20%) in the conversion of electric to visible light energy. Fluorescent lamps have been widely used both in internally and externally illuminated designs (Bovee et al., 1962; Krall and Kok, 1960; Wallman and Dodson, 1962; Ward et al., 1963). These systems are characterized by many lamps, low power input (kw./sq. m.), low yield (gm. algae/sq. m.) and relatively high efficiency (gm. algae/kw.-hr.).

The Quartzline lamp is small, shorter lived (2,000 hr.), less efficient (5 to 10%, depending on operating voltage), and has a very high surface illuminance (> 20,000 ft.-c.). Quartzline lamps have been used primarily in an effort to achieve high algal (oxygen) yield with minimum culture and system volume. In practice the lamp is usually placed inside a cooling jacket and submerged in the culture vessel. Exchange systems employing the Quartzline are characterized by few lamps, high power input, high yield, and low efficiency.

While the need for continuous and constant illumination has largely dictated the use of artificial light sources, sunlight illumination remains an attractive (if not mandatory) choice for application in space. Solar illumination has been used in numerous studies on the mass culture of algae for food and other purposes (Mayer et al., 1964; Myers et al., 1951; Tamiya, 1957). The feasibility of using near continuous sunlight for photosynthetic gas exchange has been demonstrated by Gafford and Fulton (1962). The use of direct solar energy in space has the obvious advantage that no power generation equipment is required. However, this advantage is partially offset by the need for deployment, orientation, and protection of large collecting surfaces.

Culture vessels

Features of vessels for algal culture have been given by Tamiya (1957). Although differing greatly in appearance and other characteristics, they may be divided, for purposes of discussion, into types externally and internally illuminated. Externally illuminated vessels include panels of varying thickness (Bovee et al., 1962; Gafford and Fulton, 1962; Ward et al., 1963), cylinders (Cook, 1951), and hemispherical domes (Gafford and Fulton, 1962). Internal illumination has been employed in rectangular tanks (Krall and Kok, 1960; Wallman and Dodson, 1963), cylindrical tanks (Hannan and Patouillet, 1963; Hannan et al., 1963; Matthern and Koch, 1964), and various annuli (Eley et al., 1964; Gaucher et al., 1960b; Shuler, 1963; Tew et al., 1962). Materials of construction include several types of plastic and glass and stainless steel. A review of the materials of construction and their effect on algal growth has been given by Dyer and Richardson (1962).

From the standpoint of performance, the essential feature of any algal culture vessel is the illuminated area-to-culture volume ratio. For a given irradiance, it is known that yield (gm. algae/day) is directly proportional to the illuminated area and independent of culture depth (Bovee *et al.*, 1962; Gafford and Fulton, 1962; Tamiya *et al.*, 1953; Ward *et al.*, 1963). This principle has become amply demonstrated using flat culture vessels with well-defined surface areas exposed to external illumination. The situation may be more complex in culture vessels illuminated internally. Data on surface area and surface-to-volume ratio for the various designs given in Table 2 are calculated on the basis of area of culture exposed to *incident* irradiance. This method introduces a bias in comparing yields per unit area of internally and externally illuminated designs. For the internally illuminated vessels the surface area is in all cases the cylindrical surface of the lamp or the lamp housing, whichever applies. A more correct procedure might be to define exposed area on the basis of *compensating* irradiance rather than incident irradiance. The argument for this is that any irradiance greater than that required to balance basal metabolism is photosynthetically useful. For a cylindrical light source surrounded by culture, the compensating irradiance is seen at a greater distance from the lamp than the incident irradiance. Hence, the corresponding exposed area is also greater. In a panel design illuminated from the outside, the exposed area is the same on either basis. When yield is expressed per unit area, this factor introduces a bias in favor of internal illumination which may be significant when the irradiance is very high. In spite of the foregoing argument no attempt was made to calculate compensation point surface areas owing to the difficulties involved.

A diffusing cone technique has been used to attenuate high irradiance and is also an effective method of increasing the surface-to-volume ratio in a culture vessel (Mayer *et al.*, 1964; Myers and Graham, 1961). In the experiments of Myers and Graham (1961), a 10:1 attenuation of surface area produced a twofold increase in yield for the same energy input. The diffusing cone principle may be useful for attenuation of intense sunlight in space for bioregenerative application.

Phase contacting and separation

The most widely used method for phase (liquid-gas) contacting in present algal exchangers is the bubbling or sparging technique; carbon dioxide-enriched gas is admitted under pressure through small orifices or tubes in the bottom of the culture vessel (Cook, 1951; Leone *et al.*, 1963; Zuraw *et al.*, 1961) or a separate gassing tower (Wallman and Dodson, 1962; Ward *et al.*, 1963). Phase separation is effected by venting the displaced and undissolved gases off the top of the culture. While this technique is expedient, it has several disadvantages. For efficient transfer of carbon dioxide, the entering gas must be forced through minute orifices or sintered plates and broken up in the culture by agitation. Both of these functions require power. In addition, plates tend to become fouled with algae in extended operation. Also, the method is not adaptable to a gravity-free environment.

These difficulties are in part overcome by the venturi contractor of Bauer *et al.*, (1963). The algal culture is recirculated through an external loop containing the contactor. Culture flow through the venturi creates suction at the throat, where gas is admitted and dispersed. While this technique requires a recirculating culture it provides excellent phase contact and is well adapted to a gravity-free environment. Gas separation in the absence of gravity could then be effected by a centrifugal liquid-gas separator such as that developed by Wallman *et al.* (1962).

Another potential gravity-independent method for phase contact and separation utilizes the principle of diffusion through semipermeable membranes. Attempts at practical application of this method have been largely unsuccessful (Gafford and Richardson, 1960; Newland and Price, 1963). The requirement for large membrane surfaces due to low oxygen permeability appears to be the major limiting factor.

Circulation (mixing)

Circulation or mixing of algal cultures is essential to keep the cells in suspension, in contact with dissolved nutrients and exposed to light. For long-term growth this is more than a casual problem. Algae, as well as other microorganisms, will tenaciously adhere to surfaces not exposed to vigorous agitation.

Another purpose of mixing in dense algal suspensions may be to increase the efficiency of light utilization. It is well known that algae can use high-intensity light efficiently if the energy is presented in short flashes separated by longer dark periods (Kok, 1956; Phillips and Myers, 1954). Presumably one could take advantage of this fact in dense, highly agitated cultures of algae, provided the proper time scale for turbulent mixing could be established. While yield benefits from increased turbulence have been observed (Davis, 1953; Miller *et al.*, 1964), it is not clear that efficiency is measurably improved when the power required to generate the turbulence is included.

Control systems

Of the many variables important to algal growth, only temperature, culture density and to a lesser extent, pH, have been successfully subjected to automatic control.

The optimal temperature range for most algae is sufficiently broad that temperature control in experimental equipment is no problem. In a larger sense, however, the situation in space may be different. The inherent inefficiency of photosynthetic systems may be attended by a formidable energy removal problem. If artificial light is used, a minimum of 96% of the input illumination power must be removed from the culture as heat. While it is possible that this energy can be used for other purposes, heat dissipation must be recognized as a potential future problem.

The control of population density is important to obtain the maximum productivity from any culture system. Growth and photosynthesis of algae submit to the theory of the "optimum catch" (Ketchum et al., 1949); that is, for any illumination system (light source and culture vessel combination) there is an optimum density for maximum production of algae (Cook, 1951; Matthern and Koch, 1964; Myers and Graham, 1959). In practice, the control of population density is accomplished by three methods: (1) periodic (e.g., daily) harvest and dilution (Leone, 1961); (2) continuous dilution and concurrent harvest (Matthern and Koch, 1964; Zuraw et al., 1960); and (3) automatic dilution by photoelectric control (Myers and Clark, 1944). Method 1 necessarily gives control over a population density range. Method 2 follows the theory of the chemostat (Novick, 1955) in which light is the limiting ingredient. This method requires a continuous and constant light input as well as careful attention to ensure against other limiting factors. Even then, unexplained fluctuations in long-term growth are likely to cause variations in population density (Krauss, 1964). Method 3 appears to be the most rational approach to constant density control. With this technique, light conditions within the culture are maintained constant, and changes in growth rate are readily observed as changes in rate of harvest or overflow.

Performance

Performance of algal gas exchangers is measured both by yield and by efficiency. In Table 2, yield is expressed as dry algae production rate per unit area and per unit volume of culture. Of these two methods, yield per unit area is a more basic criterion because it is primarily a function of irradiance or power input. Yield per unit volume is a function both of irradiance and surface-to-volume ratio (culture depth). Efficiency is expressed as dry algae production per unit energy supplied to the illumination source.

In judging performance of the various exchangers, several qualifications must be made. (1) The yield shown for each exchanger was the maximum stated production rate for the power input listed. In reports where production rate was given for several levels of electrical power, our choice was either the maximum yield or a representative yield as stated by the author. Our intent was to present each exchanger in the best possible light. (2) The performance data represent a wide variation in operating conditions, and it was not always clear that the other important variables, for example, population density, carbon dioxide concentration, etc., were necessarily optimal. (3) Discretion should be exercised in comparing data obtained in batch growth with data obtained in continuous or semicontinuous culture. It is well known that continuous culture yield predictions based on batch growth data are usually optimistic. (4) To enable comparison it was necessary to convert from other methods of expression to our chosen basis of dry algae production. Yields given in oxygen production were converted assuming a 1 liter oxygen to 1 gm. dry algae equivalence. Yields given as wet algae production were converted assuming a 75% moisture content of wet (packed) algae. (5) The value shown for illumination power input is for most of the exchangers a result of our calculation. Where power input was not stated, it was calculated from the number of lamps and the lamp wattage. For fluorescent lamps, the rated lamp wattage was used and the ballast losses neglected. For Quartzline lamps operated at reduced voltage, the wattage was obtained from a power versus voltage curve supplied by the General Electric Company.

Because of these limitations, detailed performance contrast between the various exchangers is perhaps presumptuous and of questionable value. *In toto*, however, these systems represent the "state of the art" in photosynthetic gas exchanger development and it is instructive to examine design criteria from the standpoint of over-all performance.

The most important factor affecting exchanger performance is irradiance, which, for engineering purposes, may be assumed equivalent to electrical power input. This effect is shown in Figure 3 where yield and efficiency are plotted as functions of illumination power input. The data represent the various exchanger systems in Table 2. For comparison and to aid in interpretation, the corresponding relations predicted from the Bush equation are also included for a hypothetical 20% efficient lamp. Quantitative agreement between theory and these data could not be expected and is largely precluded by the variation in results. Qualitatively, how-

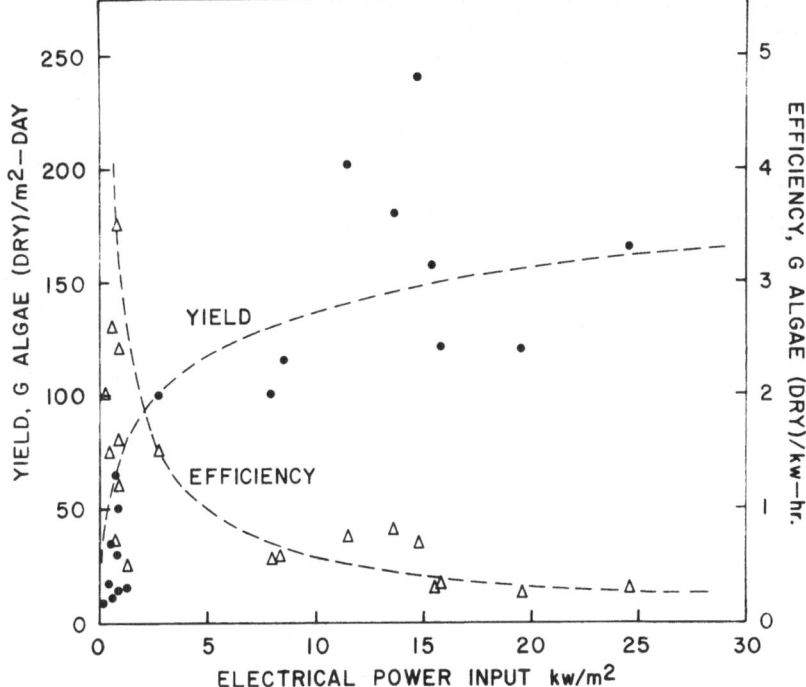

Figure 3. Yield per unit area and efficiency, gm. algae/kw.-hr.,
as a function of electrical power input for illumination, kw. Data are
for photosynthetic exchangers listed in Table 2. ●—yields; △—efficiencies. Curves calculated from Bush equation for a hypothetical lamp
20% efficient in converting electrical to visible light energy. Parameters in the Bush equation were assigned these values: $E = 0.20$; I
$= 3.5$ mw./sq. cm.; $J = 5.5$ kcal./gm. algae.

ever, the curves show the expected effect which is largely verified
by the data. It is clear that yield per unit area can be increased by
increasing *power input* but only with a concurrent and severe reduction in efficiency.

It is of interest to compare the highest efficiencies shown in
Figure 3 with the maximum theoretically attainable. Assuming a
maximum (fluorescent lamp) efficiency of 20% for the conversion of
electrical to visible light energy, and a maximum algal efficiency
of about 20% for the conversion of visible light to chemical energy,
the maximum attainable over-all efficiency becomes 4% or about
6.3 gm. dry algae/kw.-hr. In Figure 3 the highest efficiencies
are about half this maximum value. However, it must be stated that
maximum efficiency has not been a specific objective.

Figure 4 attempts to illustrate the functional dependence of yield per unit volume on surface-to-volume ratio. From theory, represented by the Bush equation, one would expect to observe a linear relationship at constant irradiance as indicated. To approximate constant irradiance, the exchanger data are grouped according to fluorescent and quartzline sources. This approximation is admittedly crude, but serves to confirm the predicted dependence. For each type of lamp, yield per unit volume is approximately a linear function of surface-to-volume ratio.

The main purpose in presenting the foregoing relationships is to confirm the importance of design criteria that have been stated by others. It may be profitable to summarize what is believed to obtain for an optimum design. First, regardless of light source, it is imperative to provide a large illuminated surface area to obtain

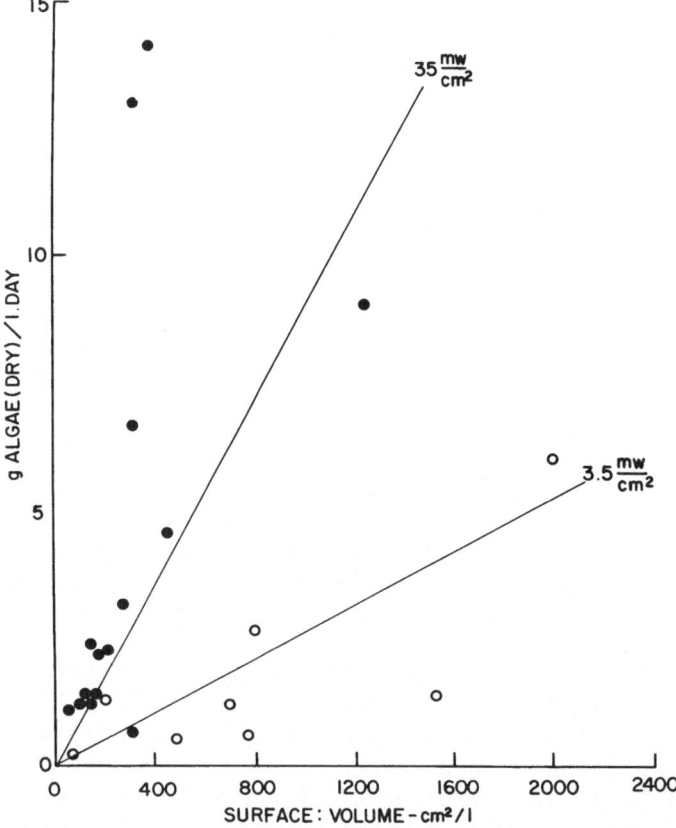

Figure 4. Yield per unit volume as a function of surface to volume ratio. Data are for photosynthetic exchangers in Table 2. ○—fluorescent lamp sources; ●—Quartzline sources. Curves were calculated from the Bush equation with values assigned in Figure 3. Parameter is incident irradiance.

maximum output. Second, minimum illuminated culture depth is requisite to maximum yield per unit volume and hence minimum total culture volume. Third, irradiance and hence light source and power input are necessarily related to area and volume, and compromise must ultimately be made for any intended application.

Figure 5 shows approximately the compromise that is necessary between area and power input for a one-man support application. The data are the extrapolated area and power required for the exchangers in Table 2 based on a one-man requirement of 600 liters oxygen per day (equivalent to 600 gm. dry algae per day). The curves represent the Bush equation for two hypothetical lamps of the stated efficiencies. Here, as before, it was not our intent to "fit" the data with theory, but only to indicate the expected relationship. It is apparent that power and area are inversely related. A large exposed surface is necessary to minimize power. Conversely, area and volume may be reduced by increasing power input, but only with attendant loss in efficiency and increased need for heat dissipation.

Perhaps as well as anything, Figure 5 presents the current

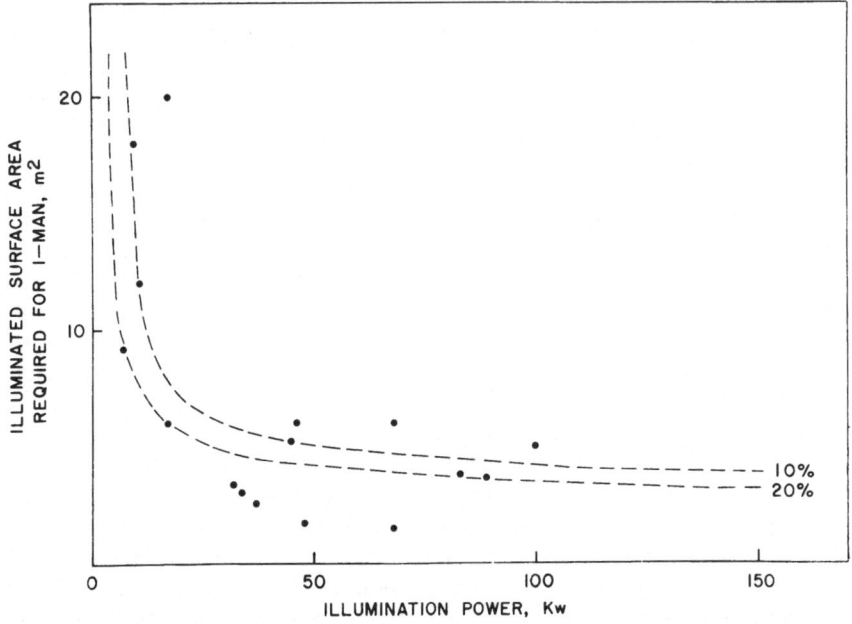

Figure 5. Illuminated surface area required for one-man support as a function of illumination power. Data are for photosynthetic exchangers in Table 2. Curves were calculated from Bush equation with values assigned in Figure 3. Parameter is hypothetical efficiency for electrical to visible light energy conversion.

status in the development of photosynthetic gas exchangers. It is evident that significant improvement in design and performance is both possible and necessary.

FUTURE DEVELOPMENT

A considerable literature, mostly in the form of laboratory reports, has developed on algal gas exchange during the past 14 years. However, the design characteristics and logistics of operation of a man-rated algal exchanger have yet to be adequately defined. The justification for continued effort on algal systems is that there is no proven alternative for missions of long duration. Development of algal exchangers is still an active area of investigation despite a recent report to the contrary (Simons, 1964), and another describing the photosynthetic system as a "technologic will-o-the-wisp" (Neswald, 1965).

The time for philosophical testifications to the efficacy of algae (or any system) for regeneration has passed. Feasibility has been demonstrated and study of laboratory models has provided useful experience. However, translation of most of the information available to design of man-supportive systems is questionable. Estimations are made only within wide limits. Consequently, systems analyses based on presently available data are premature and of doubtful value. Solid experimental data, obtained by extensive operation of prototype systems, are needed for accurate logistic comparisons. While systems engineering is greatly needed, a number of associated problem areas, such as AQ-RQ balance (Eley and Myers, 1964), medium reconstitution (Krauss and Thomas, 1954), recovery of plant nutrients from wastes (Brown et al., 1964; Golueke and Oswald, 1964), and acceptability and food value of algae (Krauss, 1962), need to be resolved before final evaluation of algae for space application. There are no known principles operating to prevent their solution.

Animal and human life-support experiments with algal exchangers have had varied degrees of success (Bovee et al., 1962; Bowman and Thomae, 1960, 1961; Eley and Myers, 1964; Golueke and Oswald, 1963; Zuraw, 1961; Zuraw et al., 1960). Because of gas leaks, much of the work is of doubtful value. Time periods of operation have been sufficient to allow confidence in reliability and dispel once prevalent fears of buildup of toxic gases (Gafford et al., 1958; Wilks, 1959). Leak problems encountered emphasize the need for more sophisticated engineering in the development of prototype exchangers.

Foaming and sticking are also problem areas and should be

submitted to engineering analysis. Proper mixing generally prevents sticking, and foaming is believed to be associated with algal excretory products and is probably influenced by the resident bacterial flora. The use of antifoam agents for foam control seems to admit defeat.

Design of algal systems has lacked the advantage of having a completely definable application and has been confused by the multitude of limitations (especially power) imposed by criteria for space application. A planetary base would provide a sound framework for which the capability of an algal system could be established. The power requirement for an algal system is high but of secondary importance if credence is given to Edward Teller's statement (1964):

> Nuclear energy provides a very ample supply of energy, and it is entirely permissible—particularly in a fixed base—to use this energy wastefully. In fact, practically the only problem is how to get rid of the energy, or to cool it, once it is created; therefore, when designing the intricate, important, and, hopefully, light units for food synthesis, I think it would be well if it is assumed that as much energy as desired can be obtained. Don't economize on energy. Economize on weight.

Nonetheless, attainment of theoretical electrical efficiency with reasonable weight and volume requirement is desirable. Significant improvement over theoretical values will require major improvement in the conversion of electrical energy into light or conversion of light energy into chemical energy by the green plant.

Presently available data are sufficient to preclude the use of algae for mission times less than several months. But algae are a best choice as one of the biological components of systems where near complete balance is desired (Johnson and Finn, 1963). The possibility of a compact, efficient, algal system is remote, but weight, volume, and power flexibility may ultimately be of advantage in the design of life-support systems for specific space missions, provided long-term reliability can be demonstrated.

REFERENCES

Acker, J. E.., and J. A. Stern (1960). System analysis of four methods of achieving closed circuit respiratory support. *In* Closed circuit respiratory systems symp. USAF WADD Tech. Rept. 60-574, W-P AFB, Ohio.

Anonymous (1956). Conference on photosynthetic gas exchangers. Off. Naval Res. Symp. Rept. ACR-13, Wash., D. C.

_____ (1960). Closed circuit respiratory systems symp. USAF WADD Tech. Rept. 60-574, W-P AFB, Ohio.

_____ (1965a). Growing food in a spaceship (closed system). Bibliography No. 1015, NASA, Wash., D. C.

_____ (1965b). Soviet bioastronautics and manned space flight. *In* Surveys of Soviet-bloc scientific and technical literature. ATD Rept. P-65-14, Library of Congress, Wash., D. C.

Arnon, D. I. (1961). Energy conversion in photosynthesis. *In* Medical and biological aspects of the energies of space (P. A. Campbell, ed.). Columbia Univ. Press, New York.

_____ (1962). Newer aspects of subcellular photosynthesis. *In* Lectures in aerospace medicine, Brooks AFB, Texas.

Barnothy, M. F. (1962). First biomagnetic symposium. Nature, 193:1243-1244.

Bassham, J. A. (1954). Use of controlled photosynthesis for maintenance of gaseous environments. Rept. UCRL-2707, Univ. Calif. Rad. Lab. Contract No. W-7405-eng-48, Berkeley, Calif.

Bates, J. H. (1961). Recent aspects in the development of a closed ecological system. Aerospace Med., 32:12-24.

Bauer, W. G., A. G. Fredrickson, and H. M. Tsuchiya (1963). Mass transfer characteristics of a venturi liquid-gas contactor. Ind. Eng. Chem. Process Design Develop., 2:178-187.

Beischer, D. E. (1963a). Biological effects of magnetic fields in space travel. *In* Proc. 12th Intern. Astronaut. Congr. (R. M. L. Baker, Jr. and M. W. Makemson, eds.). Vol. 2, pp. 515-525. Academic Press, New York.

_____ (1963b). Biomagnetics. *In* Lectures in aerospace medicine, Brooks AFB, Texas.

Benoit, R. J. (1964). Mass culture of microalgae for photosynthetic gas exchange. *In* Jackson, D. F., pp. 413-425.

_____, R. Trainor, and A. Bialecki (1960). Selection of an alga for a photosynthetic gas exchanger. USAF Rept. WADD-TDR-60-163, W-P AFB, Ohio.

Beyers, R. J. (1963). Balanced aquatic microcosms—their implications for space travel. Am. Biol. Teacher, 25:422-429.

Blasco, R. J. (1965). Nature and role of bacterial contaminants in mass cultures of thermophilic *Chlorella pyrenoidosa*. Appl. Microbiol., 13:473-477.

Bongers, L. (1964). Sustaining life in space—a new approach. Aerospace Med., 35:139-144.

_____, and B. Kok (1964). Life support systems for space missions. Develop. Ind. Microbiol., 5:183-195.

Bovee, H. H., A. J. Pilgrim, L. S. Sun, J. E. Schubert, T. L. Eng, and B. J. Benishek (1962). Large algal systems. *In* Robinette, J., pp. 8-19.

Bowman, N. J. (1953). The food and atmosphere control problem in space vessels. P. II. The use of algae for food and atmosphere control. J. Brit. Interplanet. Soc., 12:159-167.

Bowman, R. O., and F. W. Thomae (1960). An algal life support system. Aerospace Eng., 19:26-30.

_____, and F. W. Thomae (1961). Long-term non-toxic support of animal life with algae. Science, 134:55-56.

Brown, L. R., M. V. Kennedy, and R. G. Tischer (1964). An algal medium produced from human wastes. Develop. Ind. Microbiol., 6:245-249.

Brunel, J., G. W. Prescott, and L. H. Tiffany, eds. (1950). The culturing of algae. Charles F. Kettering Foundation, Yellow Springs, Ohio.

Burk, D., G. Hobby, and T. A. Gaucher (1962). Closed-cycle air purification with algae. In Man's dependence on the earthly atmosphere (K. E. Schaefer, ed.). Macmillan, New York, and USAF Rept. AMRL-TDR-63-121 91958), W-P AFB, Ohio.

Burlew, J. S., ed. (1953). Algal culture from laboratory to pilot plant. Carnegie Inst. Wash. Publ. 600.

Carpenter, E. F., ed. (1955). Trans. conf. use solar energy, 4 (Photochemical processes). Univ. Arizona Press, Tucson.

Casey, R. P., J. A. Lubitz, R. J. Benoit, B. J. Weissman, and H. Chau (1963). Mass culture of Chlorella. Food Technol., 17:85-89.

Clamann, H. G. (1959). Some metabolic problems of space flight. Federation Proc., 18:1249-1255.

_____(1961). Energy conversion and balance in manned space vehicles. In Medical and biological aspects of the energies of space (P. A. Campbell, ed.). Columbia Univ. Press, New York.

_____(1963). Chemical versus biologic recovery and recycling processes. In Lectures in aerospace medicine, Brooks AFB, Texas.

Cook, P. M. (1951). Chemical engineering problems in large scale culture of algae. Ind. Eng. Chem., 43:2385-2389.

Davis, E. A. (1953). Turbulence. In Burlew, J. S., pp. 135-138.

_____, J. Myers, and J. Dedrick (1953). Quantitative studies in controlled experimental culture units, carbon dioxide concentration. In Burlew, J. S., pp. 117-119.

Davis, L. D., K. Pappajohn, and I. M. Plavnieks (1962). Bibliography of the biological effects of magnetic fields. Federation Proc. 21, Suppl. 12.

Dyer, D. L. (1963). Regeneration: chemical or biological. In Proc. 12th Intern. Astronaut. Congr. (R. M. L. Baker, Jr. and M. W. Makemson, eds.). Vol. 2, pp. 501-514.

_____, and R. D. Gafford (1961). Some characteristics of a thermophilic blue-green alga. Science, 134:616-617.

_____, and R. D. Gafford (1962). The use of *Synechococcus lividus* in photosynthetic gas exchangers. Develop. Ind. Microbiol., 3:87-97.

_____, and D. E. Richardson (1962). Materials of construction in algal culture. Appl. Microbiol., 10:129-131.

_____, and S. Wildman (1964). Development of a balanced stock concentrate for algal medium. Develop. Ind. Microbiol., 5:365-372.

_____, D. E. Richardson, and R. P. Vandenberg (1963). Design, fabrication and test of a solar-illuminated photosynthetic gas exchanger. Final Rept. Contract AF41(609)-1606, Martin Co., Denver, Colo.

Eley, J. H., Jr., and Jack Myers (1964). Study of photosynthetic gas exchange, a quantitative repetition of the Priestley Experiment. Texas J. Sci., 26:296-333; and USAF Rept. SAM-TDR-64-52, Brooks AFB, Texas.

Enebo, L. (1960). On the supply of oxygen and food during longlasting space journeys. Astronautik (Sweden), 2:103-126.

Eppley, R. W., F. M. MaciasR, and D. L. Dyer (1964). Evaluation of certain marine algal flagellates for mass culture. USAF Rept. SAM-TDR-63-91, Brooks AFB, Texas.

_____, D. L. Dyer, and F. M. MaciasR (1964). Growth and culture characteristics of certain marine algal flagellates for mass culture. USAF Rept. SAM-TDR-64-63, Brooks AFB, Texas.

Eyster, H. C. (1964). Micronutrient requirements for green plants, especially algae. *In* Jackson, D. F., pp. 86-119.

_____, (1965). Mineral nutrient requirements of Chlorella in a continuous pure culture. First Ann. Rept. on Contract AF41(609)-2414, Monsanto Res. Corp., Dayton, Ohio.

Felfoldy, L. J. M. (1964). Experiments to select strains for algal mass culture. Ann. Biol. Tihany (Hungary), 31:177-183.

Fogg, G. E. (1953). The metabolism of algae. Wiley, New York.

Fredrickson, A. G., A. H. Brown, R. L. Miller, and H. M. Tsuchiya (1961). Optimum conditions for photosynthesis in optically dense cultures of algae. Am. Rocket Soc. J., 31:1429-1435.

Gafford, R. D. (1962). Algal photosynthetic gas exchange systems. Yale Sci. Mag., Dec.

_____ (1963). Algal photosynthetic gas exchangers. Proc. Lunar Planet. Exploration Colloq., 3:111-117, North American Aviation, Inc., Downey, Calif.

_____, and R. V. Bailey (1963). Flight test of a gravity-independent photosynthetic gas exchanger, a feasibility study. USAF Rept. SSD-TDR-63-240, Space Systems Division, Los Angeles, Calif.

_____, and C. E. Craft (1959). A photosynthetic gas exchanger capable of providing for the respiratory requirement of small animals. USAF SAM Rept. 58-124, Brooks AFB, Texas.

_____, and J. D. Fulton (1962). Solar illuminated photosynthetic gas exchangers. In Robinette, J., pp. 42-63, and USAF Rept. AAL-TDR-62-36, Fort Wainwright, Alaska.

_____, and D. E. Richardson (1960). Mass algal culture in space operations. J. Biochem. Microbiol. Technol. Eng., 2:299-311.

Galloway, R. A., and R. W. Krauss (1959). The differential action of chemical agents, especially polymyxin-B, on certain algae, bacteria, and fungi. Am. J. Botany, 46:40-49.

_____, and R. W. Krauss (1961). The effect of CO_2 on pH in culture media for algae. Plant Cell Physiol. (Tokyo), 2:331-337.

Garrick, L. S. (1963). The status of development of photosynthetic gas exchangers. NASA Rept., Langley Res. Center, Langley Station, Hampton, Va.

Gaucher, T. A., R. J. Benoit, D. Kippax, and A. Bialecki (1960a). Feasibility study on photosynthetic gas exchange. Rept. SPD 60-021, General Dynamics/Electric Boat, Groton, Conn.

_____, R. J. Benoit, and A. Bialecki (1960). Mass propagation of algae for photosynthetic gas exchange. J. Biochem. Microbiol. Technol. Eng., 2:339-359.

Gaume, J. G. (1957). Plants as a means of balancing a closed ecological system. J. Astronautics, 4:72-75.

Gini, B. (1960). A closed ecological system for extended travel, a review of pertinent literature. Library Bull. No 2, U. S. Army Quartermaster Food and Container Inst., Chicago, Ill.

Godward, M. B. E. (1962). Invisible radiations. In Lewin, R. A., pp. 551-566.

Golueke, C. G. (1962). The use of photosynthesis in the control of enclosed environments. Am. J. Public Health, 52:258-265.

_____, and W. J. Oswald (1963). Closing an ecological system consisting of a mammal, algae, and non-photosynthetic organisms. Am. Biol. Teacher, 25:522-528.

_____, and W. J. Oswald (1964). Role of plants in closed systems. Ann. Rev. Plant Physiol., 15:387-408.

_____, W. J. Oswald, and P. H. McGauhey (1959). The biological control of enclosed environments. Sewage Ind. Wastes, 31:1125-1142.

Halpern, M. H., and J. J. Konikoff (1964). Effects of magnetic fields on the growth of algae. Presented at the 35th Ann. Meeting Aerospace Med. Assoc., Miami Beach, Florida (May 11-14)

Hannan, P. J. (1964a). Effect of pressure on oxygen production by algae. Develop. Ind. Microbiol., 6:229-237.

_____ (1964b). Effect of high magnetic fields on respiration and photosynthesis in algae. Naval Res. Lab. Rept. 6153, Wash., D. C.

_____, and C. Patouillet (1963). Gas exchange with mass cultures of algae. I. Effects of light intensity and rate of carbon dioxide input on oxygen production. Appl. Microbiol., 11:446-449.

_____, C. Patouillet, and R. Shuler (1961). Studies of oxygen production by mass culture of algae. Naval Res. Lab. Rep. 5689, Wash., D. C.

_____, R. L. Shuler, and C. Patouillet (1963). A study of the feasibility of oxygen production by algae for nuclear submarines. Naval Res. Lab. Rept. 5954, Wash., D. C.

Hemerick, G. (1963). Preliminary evaluation of a lenticular algal culture vessel. In D. E. Leone et al., Chap. 4.

_____, and R. J. Benoit (1962). Engineering research on a photosynthetic gas exchanger. Pt. 1. Experimental. Rept. U413-62-018, General Dynamics/Electric Boat, Groton, Conn.

Hobby, G. L. (1959). Performance analysis of biological gas exchangers for closed ecological systems. Calif. Inst. Tech., Jet Prop. Lab. Ext. Publ. 804, Pasadena, Calif.

Ingram, W. T., ed. (1958). The engineering biotechnology of handling wastes resulting from a closed ecological system. Air Force Off. Sci. Res. Rept. No. TR 58-148, ASTIA Doc. No. AD162277, New York Univ., New York.

Jackson, D. F., ed. (1964). Algae and man. Plenum Press, New York.

Johnson, S. P., and J. C. Finn, Jr. (1963). Ecological considerations of a permanent lunar base. Am. Biol. Teacher, 25:529-535.

Kachroo, P., ed. (1960). Proc. symposium on algalogy. Indian Council Agr. Res., New Delhi.

Ketchum, B. H., L. L. Lillick, and A. Redfield (1949). The growth and optimum yields of unicellular algae in mass culture. J. Cell. Comp. Physiol., 33:267-280.

Kok, B. (1952). On the efficiency of Chlorella growth. Acta Botan. Neerl., 1:445-467.

_____ (1956). Photosynthesis in flashing light. Biochem. Biophys. Acta, 21:245-258.

_____, and L. H. Bongers (1961). Radiation tolerances in photosynthesis and consequences of excesses. In Medical and biological aspects of the energies of space (P. A. Campbell, ed.). Columbia Univ. Press, New York.

Konikoff, J. J. (1960). An engineering evaluation of algae. Rept. GE TIS R60SD469, General Electric Co., King of Prussia, Pa.

_____ (1961). A partially closed cycle life support system for long-term space flight. Ballistic Missiles Space Technol., 8:370-377.

Krall, A. R., and B. Kok (1960). Studies on algal gas exchangers with reference to space flight. Develop. Ind. Microbiol., 1:33-44.

Kratz, W. A. (1959). Photosynthetic gas exchangers and recyclers used in closed ecological system studies. In Bioastronautics, Advances in Research. Randolph AFB, Texas.

Krauss, R. W. (1958). Physiology of the fresh-water algae. Ann. Rev. Plant Physiol., 9:207-244.

_____(1962). Mass culture of algae for food and other organic compounds. Am. J. Botany, 49:425-435.

_____(1964). Discussion: Combined photosynthetic regenerative systems. *In* Conf. on nutrition in space and related waste problems. NASA Rept. SP-70, Wash., D. C.

_____, and A. Osretkar (1961). Minimum and maximum tolerances of algae to temperature and light intensity. *In* Medical and biological aspects of the energies of space (P. A. Campbell, ed.). Columbia Univ. Press, New York.

_____, and W. H. Thomas (1954). The growth and inorganic nutrition of *Scenedesmus obliquus* in mass culture. Plant Physiol., 29:205-214.

Lancaster, J. H., and R. G. Tischer (1962). Algal nutrition, a review of the literature. USAF Rept. 61-51, Brooks AFB, Texas.

Leonard, J. M. (1960). Algae and submarine habitability. Develop. Ind. Microbiol., 1:26-32.

Leone, D. E. (1961). Photosynthetic gas exchange in the closed ecosystem for space. P. II. Studies on the growth of thermophilic *Chlorella* 71105. Rept. U411-61-106, General Dynamics/Electric Boat, Groton, Conn.

_____(1963). Growth of *Chlorella pyrenoidosa* in recycled medium. Appl. Microbiol., 11:427-429.

_____, R. J. Blasco, R. J. Benoit, and G. Hemerick (1963). Growth studies on *Chlorella pyrenoidosa* 71105 and other microalgae. Rept. U413-63-031, General Dynamics/Electric Boat, Groton, Conn.

Lewin, R. A., ed. (1962). Physiology and biochemistry of algae. Academic Press, New York.

Lynch, V. H., E. C. B. Ammann, and R. M. Godding (1964). Urine as a nitrogen source for photosynthetic gas exchangers. Aerospace Med., 35:1067-1071.

Matthern, R. O., and R. B. Koch (1963). The continuous culture of algae under high light intensity. Am. Biol. Teacher, 25:502-511.

_____, and R. B. Koch (1964). Developing an unconventional food, algae, by continuous culture under high light intensity. Food Technol., 18:58-65.

Mattoni, R. H. (1963). General principles of biological regeneration for lunar base life support. Proc. Lunar Planet Exploration Colloq. 3:101-109, North American Aviation, Inc., Downey, Calif.

Mayer, A. M., U. Zuri, Y. Shain, and H. Ginzburg (1964). Problems of design and ecological considerations in mass culture of algae. Biotech. Bioeng., 6:173-190.

Meleshko, G. I. (1964). The problem of increase in the photosynthetic productivity of a *Chlorella* culture in apparatuses for
biological regeneration of air. *In* Siskayan, N. M., and V. I.
Yazdorskiy, Problems of space biology, Vol. 3. USSR Acad. Sci.
Publ. House, Moscow. Tech. trans. 64-31578, U. S. Dept. Com-
merce, Joint Publ. Res. Service, Wash., D. C.

Miller, R. L., A. G. Fredrickson, A. H. Brown, and H. M. Tsuchiya
(1964). Hydromechanical method to increase efficiency of algal
photosynthesis. Ind. Eng. Chem. Process Design Develop., 3:
134-143.

Mitchell, D. F. (1960). Genetics and the reliability of ecological
systems. *In* Ballistic missile and space technol. (D. P. Le Gal-
ley, ed.), Vol. 1, pp. 63-76. Academic Press, New York.

Moyer, J. E. (1962). Aerobic waste disposal systems. *In* Robinette,
J., pp. 281-289.

Myers, Jack (1946). Culture conditions and the development of the
photosynthetic mechanism. IV. Influence of light intensity on
photosynthetic characteristics of *Chlorella*. J. Gen. Physiol.
29:429-440.

_____ (1951). Physiology of the algae. Ann. Rev. Microbiol., 3:
157-180.

_____ (1954). Basic remarks on the use of plants as biological gas
exchangers in a closed system. J. Aviation Med., 25:407-411.

_____ (1956a). Experiments with photosynthetic gas exchangers.
In Conf. on photosynthetic gas exchangers. Off. Naval Res. Symp.
Rept. ACR-13, Wash., D. C.

_____ (1956b). Algae as an energy converter. *In* Proc. World Symp.
on Appl. Solar Energy, Stanford Res. Inst., Menlo Park, Calif.

_____ (1957). Algal cultures. *In* Encyclopedia of chemical technology. The Interscience Encyclopedia, New York.

_____ (1958). Study of a photosynthetic gas exchanger as a method
of providing for the respiratory requirements of the human in a
sealed cabin. USAF SAM Rept. 58-117, Randolph AFB, Texas.

_____ (1960a). Space logistics. II. Biosynthetic gas exchangers.
In Lectures in aerospace medicine, Brooks AFB, Texas.

_____ (1960b). The use of photosynthesis in a closed ecological
system. *In* Benson, O. O., and H. Strughold, Physics and medicine of the atmosphere and space. Wiley, New York.

_____ (1962). The algal photosynthetic gas exchanger. *In* Robinette,
J., pp. 5-7.

_____ (1963). Introductory remarks to a symposium on space biology: ecological aspects. Am. Biol. Teacher, 25:409-411.

_____ (1964a). Combined photosynthetic regenerative systems. *In*
Conf. on nutrition in space and related waste problems, NASA
Rept. SP-70, Wash., D. C.

_____ (1964b). Use of algae for support of the human in space. *In* Proc. Fourth Intern. Space Science Symp. North-Holland Publ. Co., Amsterdam.

_____, and L. B. Clark (1944). Culture conditions and the development of the photosynthetic mechanism. II. An apparatus for the continuous culture of *Chlorella*. J. Gen. Physiol., 28:103-112.

_____, and A. H. Brown (1961). Gas regeneration and food production in a closed ecological system. Natl. Acad. Sci., Natl. Res. Council Publ. 893, Wash., D. C.

_____, and C. S. French (1960). Evidences from action spectra for a specific participation of chlorophyll b in photsynthesis. J. Gen. Physiol., 43:723-736.

_____, and Jo-Ruth Graham (1959). On the mass culture of algae. P. II. Yield as a function of cell concentration under continuous sunlight irradiance. Plant Physiol., 34:345-352.

_____, and Jo-Ruth Graham (1961). On the mass culture of algae. Part III. Light diffusers-high vs low temperature *Chlorellas*. Plant Physiol., 36:342-346.

_____, J. N. Phillips, Jr., and Jo-Ruth Graham (1951). On the mass culture of algae. Plant Physiol., 26:539-548.

Neswald, R. G. (1965). Life support's new twists. Space/Aeronautics, 44:70-76.

Newland, R. G., and R. W. Price (1963). Design study of gravity-independent photosynthetic gas exchanger. USAF Rept. AMRL-TDR-63-59, W-P AFB, Ohio.

Nielsen, E. S. (1952). Experimental carbon dioxide curves in photosynthesis. Physiol. Plantarum, 5:145.

Novick, A. (1955). Growth of bacteria. Ann. Rev. Microbiol., 9:97-110.

Odum, H. T. (1963). Limits of remote ecosystems containing man. Am. Biol. Teacher, 25:429-443.

Phillips, J. N., Jr. (1961). Biological systems in space vehicles. *In* Lectures in aerospace medicine, Brooks AFB, Texas.

_____ (1962a). Experiments with photosynthetic microorganisms. *In* Crawford, G. W., Radiobiologic experiments in Discoverer satellite XVII. USAF Rept. 62-67, Brooks AFB, Texas.

_____ (1962b). Closed ecological systems for space travel and estraterrestrial habitation. Develop. Ind. Microbiol., 3:5-13.

_____, and Jack Myers (1954). Growth rate of *Chlorella* in flashing light. Plant Physiol., 29:152-161.

Popma, D. C. (1962). Life support for space stations. Astronautics, 7:44-47.

Posner, H. B., and A. H. Sparrow (1964). Survival of *Chlorella* and *Chlamydomonas* after acute and chronic gamma radiation. Radiation Botany, 4:253-257.

Robinette, J., ed. (1962). Biologistics for space systems symposium. USAF Rept. AMRL-TDR-62-116, W-P AFB, Ohio.

Rousseau, J. (1964). Atmospheric control systems for space vehicles. USAF Rept. ASD-TDR-62-527, P. II, W-P AFB, Ohio.

Semenenko, V. Ye., and M. G. Vladimirova (1963). Effect of spaceflight conditions in the satellite on the preservation of the viability of *Chlorella* cultures. *In* Sisakyan, N. M., Problems of space biology, Vol. 1. U.S.S.R. Acad. Sci. Publ. House, Moscow. NASA Tech. trans. F-174, Wash., D. C.

Shuler, R. L. (1963). Some factors affecting the oxygen production by algae in a small culture unit. Naval Res. Lab. Rept. 5979, Wash., D. C.

Simons, H. (1964). Washington microcosm. Bioscience, 14:45-46.

Sorokin, C. (1958). The effect of the past history of cells of *Chlorella* on their photosynthetic capacity. Physiol. Plantarum, 11:275-283.

_____ (1960). Photosynthetic activity in synchronized cultures of algae and its dependence on light intensity. Arch. Mikrobiol., 37:151-160.

_____, and R. W. Krauss (1962). The effects of light intensity on the growth rates of green algae. Plant Physiol., 33:109-113.

_____, and Jack Myers (1953). A high-temperature strain of *Chlorella*. Science, 117:330-331.

Specht, H. (1952). Toxicology of travel in the aeropause. *In* White, C. S., and O. O. Benson, Physics and medicine of the upper atmosphere,. Univ. of New Mexico Press, Albuquerque.

Spiegler, P. E. (1963). Bibliography of bioregenerating systems for extraterrestrial habitation. USAF Rept. AMRL-TDR-63-121, W-P AFB, Ohio.

Stambler, I. (1965). Bioscience in orbit. Space/Aeronautics, 44:46-54.

Tamiya, H. (1957). Mass culture of algae. Ann. Rev. Plant Physiol., 8:309-334.

_____, E. Hase, K. Shibata, A. Mituya, T. Iwamura, T. Nihei, and T. Sasa (1953). Kinetics of growth of *Chlorella*, with special reference to its dependence on quality of available light and on temperature. In Burlew, J. S., pp. 204-232.

Teller, E. (1964). Water generation in space. In Conf. on nutrition in space and related waste problems. NASA Rept. SP-70, Wash, D. C.

Tew, R. W., J. O. Sane, and R. P. Geckler (1962). Highly concentrated carbon dioxide as a carbon source for continuous algae cultures. *In* Robinette, J., pp. 64-80.

Thacker, D. R., and H. Babcock (1957). The mass culture of algae. J. Solar Energy Sci. Eng., 1:37-49.

Tischer, R. G., and B. P. Tischer (1963). Open sequence components of a closed ecology. Am. Biol. Teacher, 25:444-449.

Turner, J. S., and E. G. Brittain (1962). Oxygen as a factor in photosynthesis. Biol. Rev., 37:130-170.

Van Oorschot, J. L. P. (1955). Conversion of light energy in algal culture. Mededel. Landbouwhogeschool Wageningen (Nederland), 55:225-276.

Vela, G. R., C. H. Ward, and J. E. Moyer (1965). Growth of bacterial contaminants in culture of *Chlorella pyrenoidosa* TX 71105. Presented at 16th Ann. AIBS Meetings Biol. Soc., Urbana, Ill. (Aug. 15-20).

Wallman, H., and J. Dodson (1962). Design and development of an engineering model photosynthetic gas exchanger. *In* Robinette, J., pp. 20-41.

_____, J. L. Dodson, V. A. Speziali, A. E. Rabe, R. J. Nickerson, and R. R. Cordeiro (1962). Research and development of a liquid-gas contactor for photosynthetic gas exchangers. USAF Rept. AMRL-TDR-62-101, W-P AFB, Ohio.

Ward, C. H., and C. N. Guerra (1962). Growth of photosynthetic microorganisms following orbital space flight. In Prince, J. E., Biologic systems of Discoverer satellites XXIX and XXX. USAF Rept. 62-62, Brooks AFB, Texas.

_____, S. S. Wilks, and H. L. Craft (1963). Use of algae and other plants in the development of life support systems. Am. Biol. Teacher, 25:512-521.

_____, J. E. Moyer, and G. R. Vela (1964). Studies on bacteria associated with *Chlorella pyrenoidosa* TX 71105 in mass culture. Develop. Ind. Microbiol., 6:213-222 and USAF Rept. SAM-TDR-64-73, Brooks AFB, Texas.

Welch, B. E. (1961). Logistics of photosynthesis. *In* Medical and biological aspects of the energies of space (P. A. Campbell, ed.). Columbia Univ. Press, New York.

Wilks, S. S. (1959). Carbon monoxide in green plants. Science, 129:964-966.

_____(1963). Biological recycling processes in closed systems. *In* Lectures in aerospace medicine, Brooks AFB, Texas.

Zuraw, E. A. (1961). Algae-primate gas exchange in a closed gas system. Develop. Ind. Microbiol., 3:140-149.

_____, and T. E. Adamson (1963). Photosynthetic gas exchanger performance studies at high light input. Rept. U413-63-036, General Dynamics/Electric Boat, Groton, Conn.

_____, G. S. Christiansen, D. L. Kippax, R. J. Benoit, N. L. Richards, I. N. Huppert, D. E. Leone, T. A. Gaucher, R. R. Trainor, and F. F. Noe (1960). Photosynthetic gas exchange in the closed ecosystem for space. Rept. SPD 60-085, General Dynamics/Electric Boat, Groton, Conn.

_____, B. J. Weissman, R. P. Casey, V. A. Speziali, and T. A. Adamson (1961). Photosynthetic gas exchange in the closed eco-system for space. P. I. Pilot photosynthetic gas exchange studies. Rept. U411-61-131, General Dynamics/Electric Boat, Groton, Conn.

10.

CARBON DIOXIDE CONTROL BY ENZYMATIC REACTIONS IN SPACECRAFT ATMOSPHERE

G. GRAF · R. E. HOAGLAND[1]· W. R. CARL[2]· S. R. KUROWSKY[2]

Department of Chemistry, Youngstown University
Youngstown, Ohio

The feasibility of using enzymatic reactions for carbon dioxide control of closed atmospheres was investigated in two directions. The hydration and dehydration of carbon dioxide by bovine carbonate hydro-lyase (EC 4.2.1.1) was developed to a continuous laboratory scale operation. The multienzyme system, pyruvate: CO_2 ligase (EC 6.4.1.1), L-malate: NAD oxidoreductase (EC 1.1.1.37), and L-malate: NADP oidoreductase (decarboxylating) (EC 1.1.1.40), was tested for carbon dioxide exchange under static conditions.

CO_2 MANAGEMENT BY CARBONATE HYDRO-LYASE (EC 4.2.1.1)

In 1928 Henriques was puzzled by the rapid rate of the dehydration of CO_2 in the blood, which allowed 10 times as much CO_2 transfer to the lungs as was calculated without assuming enzymatic catalysis. In 1933 an enzyme, carbonic anhydrase [carbonate hydro-lyase (EC 4.2.1.1)],* designated as CA, was found in the erythrocytes (Meldrum and Roughton, 1933), and was shown to catalyze the rapid hydration and dehydration of CO_2:

$$CO_2 + H_2O \rightleftharpoons H_2CO_3 \rightleftharpoons H^+ + HCO_3^-$$

Further study (Keilin and Mann, 1940) revealed this to be the earliest known zinc metallo-enzyme. After the metal-binding and catalytic activity of carbonic anhydrase was investigated, several forms of the enzyme were differentiated in preparations from human and bovine blood (Lindskog and Nyman, 1964): Forms BCA A and BCA B for the bovine enzyme, and HCA A, HCA B, and

Present addresses: [1]Department of Chemistry, University of Cincinnati, Cincinnati, Ohio. [2]Department of Chemistry, Purdue University, Lafayette, Indiana.

HCA C for the human enzyme.

This development was followed by the determination of the amino acid composition of the enzyme (Lindskog and Nyman, 1964) and an esterase activity was demonstrated (Tashian et al., 1964).

Since 1948 carbonic anyhdrase has been frequently applied in kinetic studies involving H_2CO_3, CO_2, or HCO_3^- (Roughton, 1948). Dean Burk (1961) suggested the use of this enzyme in carbonate and amine buffers for CO_2 exchange in the Warburg manometric vessels, atomic submarines, and industrial carbon dioxide scrubbers.

The molecular weight of the enzyme is about 30,000 for all known forms. The hydration of CO_2 by carbonic anhydrase is one of the simplest and fastest enzymatic reactions. The turnover number of this enzyme is reported as 10^8/min. (Roughton and Clark, 1951). (The turnover number is defined as the number of substrate molecules reacting per active center of the enzyme per minute.) Forms HCA A and HCA B have lower specific activity than HCA C. The specific activity of the latter is nearly the same as that of BCA A and BCA B (Nyman and Lindskog, 1964). Since bovine forms do not exist on different levels of specific activity, bovine preparations yield optimum conversion rates without further need for separation.

In order to test the feasibility of CO_2 management by carbonic anhydrase three series of experiments were conducted. First, absorption data were measured manometrically in a closed system between low concentrations of CO_2 in the gas phase and a liquid absorber. Second, the absorption of CO_2 from an air stream was tested in cycling operation through a series of consecutive absorption and desorption steps. Third, a continuous absorption of CO_2 from the feed gas was achieved by circulating the absorbent liquid through an absorber and a desorber.

In all these experiments the absorbing solutions contained carbonic anhydrase from Nutrititonal Biochemicals Corp., isolated from bovine blood and purified according to the procedure of Keilin and Mann (1940). The molecular weight of the preparations was given as 31,000 and the specific activity as 1,000 units per mg. protein, that is, approximately 100 times the specific activity of bovine erythrocytes. The enzyme was suspended in a solution of the free base, 2-amino-2-(hydroxymethyl)-1, 3-propanediol, or Tris, enzyme grade, from Eastman. Nitrogen bases and their salts were considered as early as 1946 for studies of enzyme behavior. Several amine buffers are particularly suited for this purpose (Nahas, 1961). All solutions were prepared in deionized water.

Equilibrium measurements

In these experiments absorption isotherm were obtained at 25°C. for a series of buffer and enzyme concentrations by plotting the millimoles of CO_2 absorbed by 100 ml. absorbent against the partial pressure of CO_2 in a closed system.

Figure 1 is a diagram of the apparatus used for these measurements. A 1,000-ml., 4-necked flask was equipped with a 140-mm. sealed U-tube mercury manometer, a 100-ml. addition funnel graduated in milliliters, a gas inlet provided with a stopcock, and a thermometer, all of them connected to the flask through ground glass joints. The absorbent liquid in the flask was stirred by a magnetic stirrer. The CO_2 was delivered to the system from a jacketed 100-ml. gas buret graduated in 0.1 ml., which was provided with a two-way stopcock permitting the buret to be opened to a carbon dioxide cylinder or to the system, respectively. A three-way stopcock inserted between the gas buret and the system allowed for evacuation of the system. The temperature of the gas in the gas buret and both phases in the system was kept constant by equilibrating with the room temperature or by using a Haake constant temperature circulator.

The gas buret was filled from Matheson lecture bottles containing commercial grade CO_2 with a minimum purity of 99.5%. Buffers were prepared according to the procedures given by Gomori (1955).

Before making the runs, the total volume of the empty system, that is, the volume of the reaction flask plus the volume of the connections to the stopcock of the gas buret, was determined by introducing known volumes of CO_2, measured at known temperatures and pressures, into the completely dry evacuated flask. Upon each addition, the temperature and pressure of the gas in the system was recorded. From these data, the empty system volume was calculated, assuming ideal behavior. The experimental densities of CO_2 in the pressure range 0 to 40 mm. Hg, used in these experiments, agree with the densities calculated from the ideal gas law, within the precision of the measurement. Hence, the number of CO_2-millimoles delivered by each addition from the gas buret to the system is calculated as $n'_g = P'_g V'_g / RT'$. (Gas buret parameters are primed and system parameters are non-primed quantities. The subscript g stands for the gaseous phase; the subscript l for the liquid phase; n is the number of millimoles; P, pressure in mm. Hg; V, volume in ml.; T, temperature in °K.; R, 62.361 (ml.) (mm. Hg)/(millimoles) (deg. abs.); Δn, increment of mass in millimoles; Σn, accumulated mass in millimoles.) Since there is no liquid phase in the system, $n_l =$

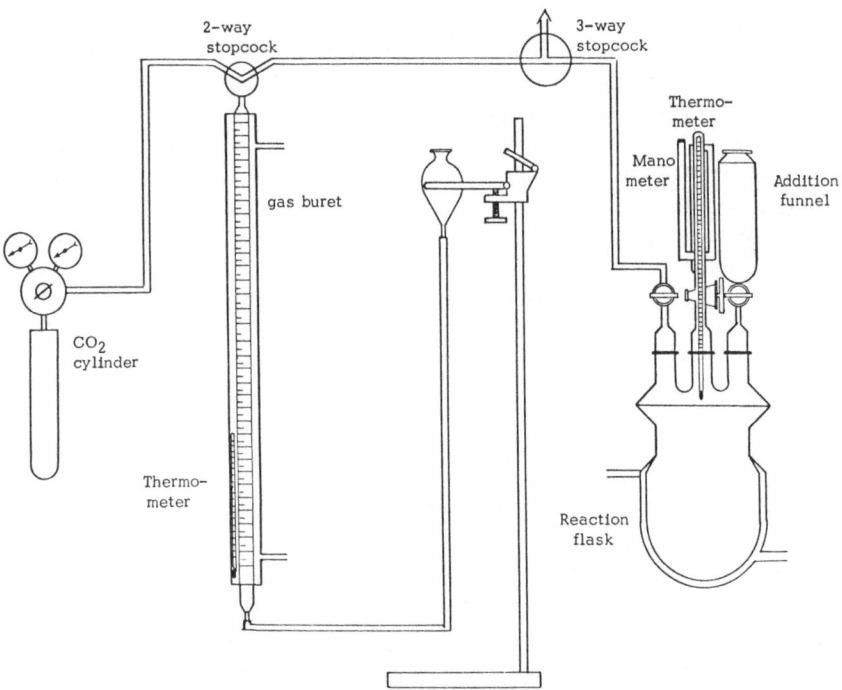

Figure 1. Manometric apparatus for measurement of CO_2 absorption.

0, and $n_g = n'_g$, and since all CO_2 is accumulated in the gas phase, $\Delta n'_g = n'_g$ = millimoles CO_2 present in the gas phase. The total empty system volume is $V = V_g$. This may be calculated from $P_{\Delta n_g}$ values, that is, the small pressure variations caused by individual CO_2 increments in the gas phase, by setting $V = n_g T_g R / P_{\Delta n_g}$, or more conveniently, by averaging over a wider range. If n_g is plotted against P_{n_g}, a straight line is obtained. Then, averaging between points (2) and (1), the volume is calculated as:

$$V = 62.361 \frac{n_g^{(2)} - n_g^{(1)}}{P_g^{(2)} - P_g^{(1)}} T_g$$

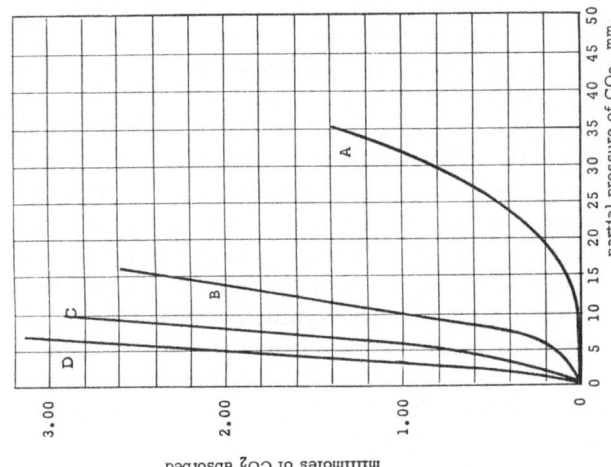

Figure 3. Effect of carbonic anhydrase on CO^2 absorption in $0.5\,M$ Tris base. Temperature: 25°C.; pH 10.80. A, 0% CA; B, 0.004% CA; C, 0.020% CA; D, 0.100% CA.

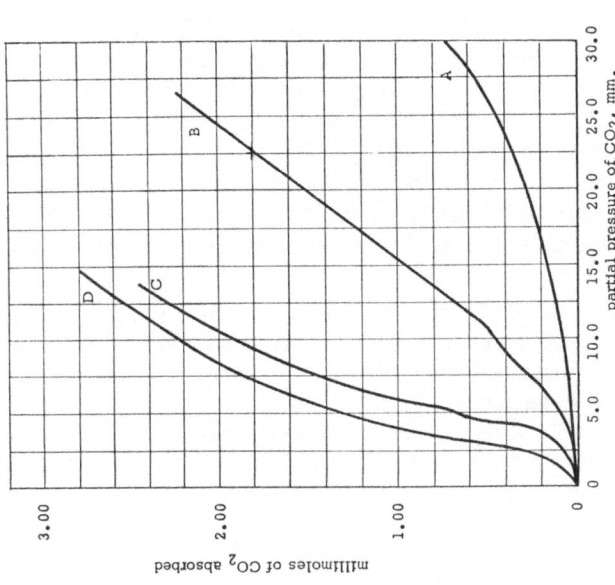

Figure 2. Effect of carbonic anhydrase on CO_2 absorption in 0.1 Tris base. Temperature: 25 C.; pH 10.40. A, 0% CA; B, 0.004% CA; C, 0.020% CA; D, 0.100% CA.

The absorption isotherm was determined by the following method: 100 ml. liquid absorbent was added to the evacuated flask through the addition funnel and the resulting pressure recorded after the liquid came to thermal equilibrium. Then known volumes of carbon dioxide at known temperature and pressure were added to the system; the liquid was stirred; and, after absorption equilibrium was reached, the pressures were recorded. From these data the millimoles of CO_2 absorbed by the liquid at the recorded pressure and temperature were calculated. If V_l ml. absorbent liquid was placed in the system, the volume of the gas phase is expressed as $V_g = V - V_l$, the number of millimoles of CO_2 in the gas phase as $n_g = P_{n_g} V_g / RT_g$, and the number of millimoles of CO_2 in the liquid phase as $n_l = \Sigma n'_g - n_g$ at P_n and T_g. T_g is the observed temperature of the gas phase, and P_{n_g} is computed by subtracting the observed initial system pressure $P_{g,0}$, that is, the vapor pressure of the liquid and the residual pressure left after evacuation of the system from the observed pressure of the system; then $P_{n_g} = P_g - P_{g,0}$. $\Sigma n'_g$ is calculated as the cumulative value of n'_g values. Each n'_g is plotted against P_{n_g} for the given temperature of the liquid absorbent.

The selection of buffer in these experiments was based on data in the literature concerning inhibitory effects on carbonic anhydrase. According to Roughton and Booth (1946), anions of many salts exhibit a general mild inhibitory effect on the bovine enzyme. This fact eliminated a large group of buffers from this study, particularly those that contain phosphate, borate, carbonate, acetate, and chloride ions. Among the various buffers used at higher concentrations, Tris · HCl buffer appeared to be the least inhibitory (Datta and Shepard, 1959). Although in these experiments several runs were made with Barbital buffer and Tris · HCl buffer, neither of them proved satisfactory. In Barbital buffers the absorption was negligible below 5 mm. Hg partial pressure of CO_2.

Figure 4. Effect of carbonic anhydrase on CO_2 absorption and desorption rate in freshly prepared solution; 0.006% CA in $0.1 M$ Tris maleate buffer, pH 7.63. *: Begin CO_2 flow at 1 atm.; **: CO_2 flow stopped.

Figure 5. Effect of carbonic anhydrase on CO_2 absorption and desorption rate in 11-day-old solution: 0.006% CA in $0.1 M$ Tris maleate buffer, pH 8.09 (stored at 5°C.). *: Begin CO_2 flow at 1 atm.; **: CO_2 flow stopped.

Figure 6. Effect on CO_2 absorption of a 19-day-old solution of carbonic anhydrase in Barbital buffer. A, $0.05 M$ Barbital buffer, pH 8.22; B, 0.01% CA in $0.05 M$ Barbital buffer, pH 8.22, 19 days old (stored at 5°C.).

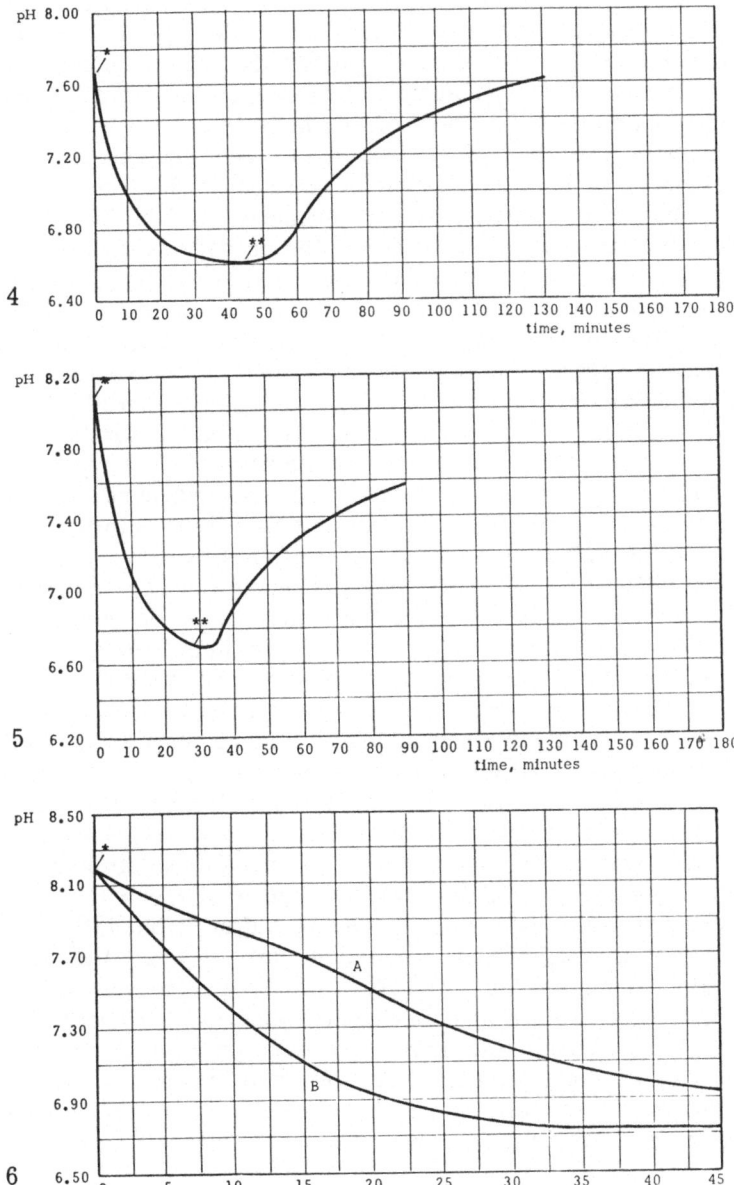

In addition to this, it was not possible to increase the buffer concentration because of the low solubility of 5,5-diethylbarbituric acid. Tris · HCl buffers contain chloride ions with inhibitory effects, which became discernible at higher concentrations. Hence, in these experiments, only solutions of the free Tris base were used, with no addition of acid. During the runs, of course, a buffering effect builds up because of the accumulation of bicarbonate and hydronium ions, products of CO_2 hydration.

Figure 2 is a typical series of runs showing the effect of increasing enzyme concentration from 0% to 0.1% in 0.1 M Tris base. The effect is considerably enhanced in a 0.5 M Tris, represented in Figure 3. In both cases, the pressure range of CO_2 was 0 to 40 mm. Hg. In the absence of enzyme, increasing buffer concentrations have only a slight effect on absorption. Addition of enzyme has a particularly enhanced effect in more concentrated solutions. Larger enzyme concentrations in concentrated Tris produce a greater tendency to absorb CO_2 at low partial pressures.

In addition to the absorption studies, semiquantitative experiments were conducted to show the stability of the enzyme and the reversibility of enzymatic activity. Decreasing partial pressures of CO_2 favor desorption. To demonstrate reversibility and stability, first buffered solutions of the enzyme were saturated by passing CO_2 at 1 atm. partial pressure through the liquid at a controlled rate and the pH change was recorded during saturation. Then, when the pH approached minimum, the gas flow was stopped and the pH change, due to desorption, was recorded again. The experiment was repeated several times with the same solution over a period of several days. Figure 4 shows the behavior of a freshly prepared 0.006% enzyme solution in 0.1 M Tris maleate buffer. Figure 5 shows the saturation and desorption pattern of the same solution after 11 days. Figure 6 presents another type of experiment which shows that a 19-day-old solution containing 0.01% enzyme in 0.05 M Barbital buffer absorbs CO_2 at a considerably higher rate than the same buffer without the enzyme. This kind of experiment formed a link between the static and dynamic technique for CO_2 absorption. Once it was established that CO_2 was rapidly absorbed and desorbed at low partial pressures by buffered solutions containing only small amounts of the enzyme, and, further, that the stability of the enzyme was reassuring, a cycling operation was set up for intermittent CO_2 absorption and desorption under dynamic conditions.

Cycling operation

In a second series of experiments CO$_2$ was intermittently absorbed from a continuous air stream of low CO$_2$ concentration and then the system was regenerated. With the same method, data were obtained concerning the capacity and some dynamic characteristics of the system in order to develop the intermittent operation into a continuous process with simultaneous absorption and desorption.

Figure 7 is a diagram of the apparatus for cycling operation. Essentially it was composed of three parts: a flow control system to regulate the flow rate and composition of the feed gas; an absorber; and a recording infrared gas analyzer to monitor the effluent gas composition.

The flow control system maintained a constant rate of feed gas of uniform composition. The components, air and carbon dioxide, were fed into the main stream from compressed gas cylinders. The air was passed through a Drierite and an Ascarite tower; the carbon dioxide, through a Drierite tube only. The component gas rates were controlled by metering valves and flowmeters. A uniform feed gas concentration was obtained by passing the gases through a premixer and a mixing tower, packed with glass beads. This gas mixture was fed to the absorber and the effluent gas was passed through the analyzer. Provisions were made to bypass the absorber for occasional direct checking of the feed gas composition with the analyzer.

The liquid absorbent was contained in a jacketed reactor equipped with gas inlet and outlet connections, thermometers, and magnetic stirring. The capacity of the jacketed glass reactor was 2,000 ml. and it had a 4-necked top provided with ground glass connections. The inlet tube was connected with an EC fritted glass dispersion plate. The termometers measured the liquid and gas temperatures within the reactor. Heating and cooling was accomplished by a combination of a Haake constant-temperature circulator and a Thermo-Cool, maintaining the temperature of the liquid during the experiments within the limits of $\pm 0.2°C$.

For the automatic and continuous analysis of the effluent gas, a Beckman infrared analyzer, Model 215, was used, sensitized for carbon dioxide and connected with a Sargent SRL potentiometric recorder. The infrared analyzer was standardized before each measurement. The zero point was set with air as zero gas, the upscale point was set with a certified standard mixture of the desired CO$_2$ concentration with balance air, from Matheson. The gain control of the analyzer and the 100% adjust of the recorder were set for full scale deflection for the CO$_2$ concentration of the feed gas. The gases were dried before entering the analyzer.

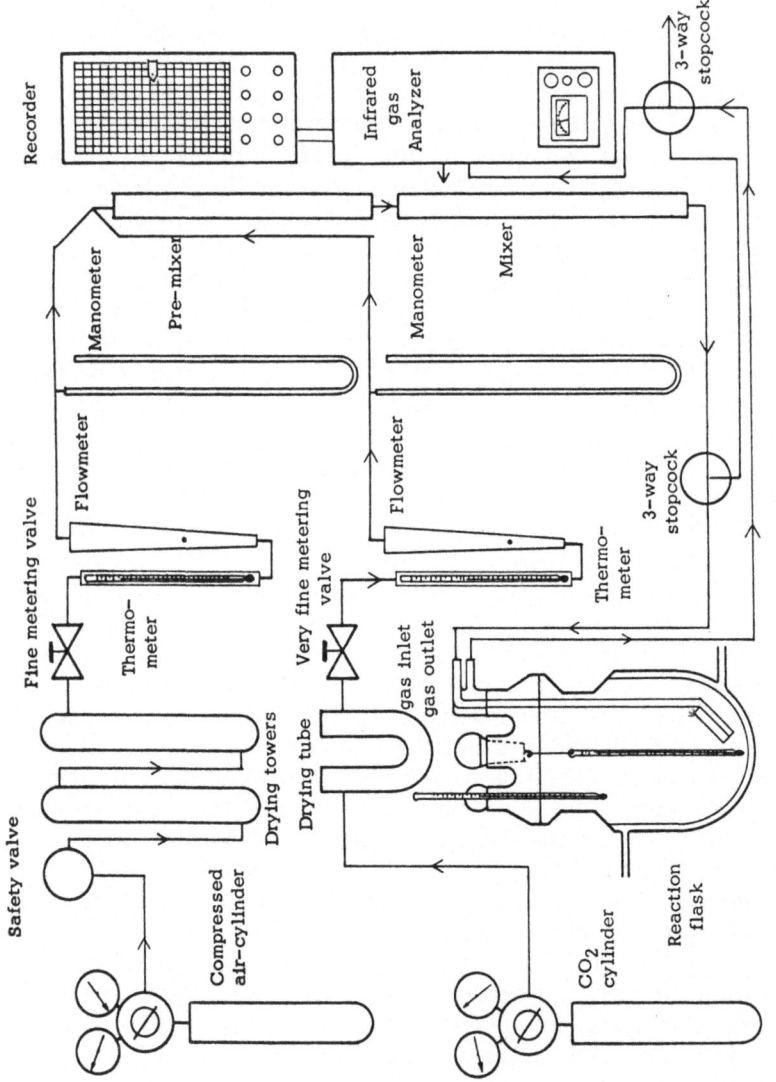

Figure 7. Apparatus for CO₂ management in cycling operation.

During calibration and the actual operation, all readings were taken from the recorder.

The chemicals and the enzyme used in these experiments were the same as those used in the study of carbon dioxide exchange under static conditions. The absorbent solutions were prepared similarly, by suspending the enzyme in Tris base solutions.

In the cycling operation, periods of 90 min. absorption were followed by periods of 60 min. desorption. Desorption was accomplished by mildly heating the liquid absorbent to 30°C. and evacuating the system simultaneously. Reduced partial pressures of CO$_2$ and slightly elevated temperatures facilitated the dehydration reaction in the presence of the enzyme. This mild treatment did not denature the enzyme and, apparently, did not impair its specific activity.

The constant-temperature bath with the circulating pump was inadequate to compensate for the rapid loss of heat due to evaporation in the vacuum. This was balanced partially by refluxing the water into the system through a condenser, but mainly by employing an infrared lamp to maintain the temperature at 30°C. The lamp was radiating at the jacketed system from a distance of approximately 1.3 cm., while the circulating heating pump was also in operation. In order to reflect the radiant heat and prevent dissipation of energy, a silver-mirrored screen was placed behind the system. This arrangement assured good control of the temperature, without local overheating and inactivation of the enzyme.

A series of intermittent absorption-regeneration runs usually required a period of 2 or 3 days. Overnight, the solutions were kept unregenerated to secure maximum substrate stabilization. Much information was gained concerning enzyme stability through these extended periods of observation on the same system.

The residual CO$_2$ in the effluent gas was recorded against time. The total volume of CO$_2$ in the feed gas at operating conditions, entering the absorber over a certain length of time, is represented by the area of the chart over the time axis, provided the full chart scale corresponds to the CO$_2$ concentration of the feed gas. The area under the recorded curve represents the volume of residual CO$_2$ at operating conditions in the effluent gas, and the corresponding area above the curve gives the volume of CO$_2$ retained by the absorber. The areas were obtained by integration. From these data, the mass flow rate of CO$_2$ in millimoles per minute, the toal millimoles of CO$_2$ entering the absorber, and the total millimoles of CO$_2$ retained by the absorber were calculated. The average rate of absorption was given in millimoles of CO$_2$ per minute averaged over the total absorption time.

During the first 10 to 15 min. of the experiment the curves

pass through a deep minimum which is followed immediately by a slight maximum. When the effluent gas begins to flow toward the analyzer, an air plug is pushed ahead in the tubing leading to the sample cell of the analyzer. This is followed by a slight accumulation of CO_2, due probably to compressibility differences of the gases. When the effluent gas starts its regular flow through the sample cell, the curve levels off to a plateau, indicating the actual CO_2 content of the effluent. The length and height of this nearly horizontal plateau indicates the efficiency of absorption. This horizontal portion of the curve is the region of steady-state performance.

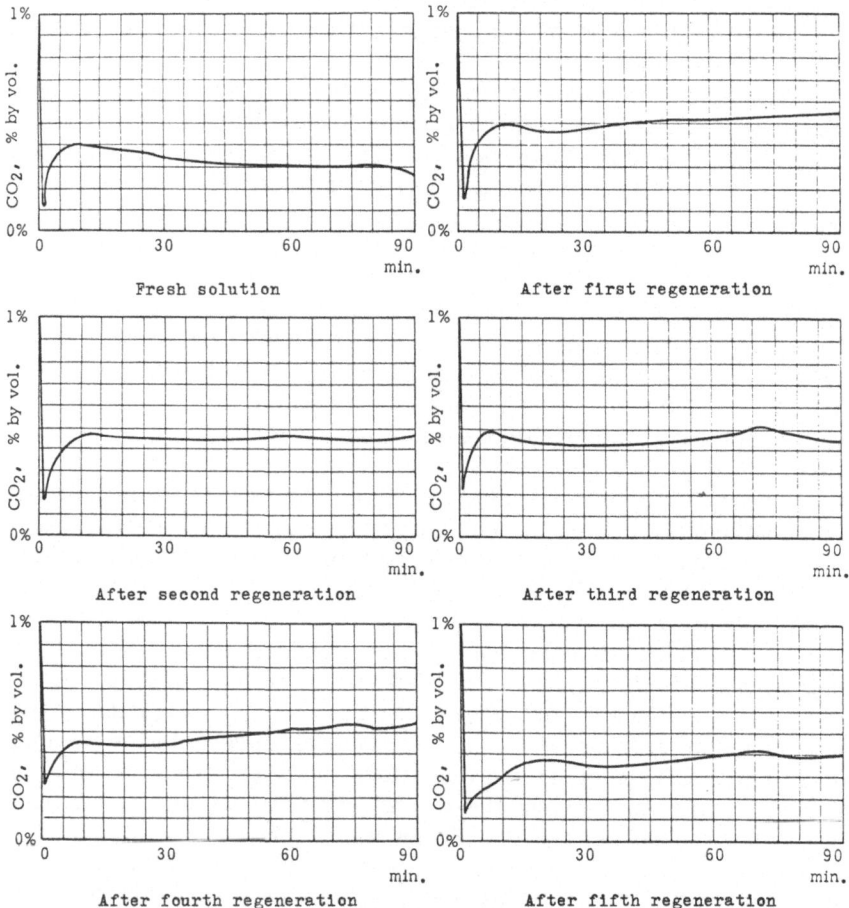

Figure 8. CO_2 absorption in cycling operation. Absorber: 0.01% CA in 0.04 M Tris, 1.000 ml.; feed gas: 1% CO_2 in air; feed rate: 500 cc./min.; temperature: 20° C. Recorded is the percent of CO_2 in effluent.

When the absorption rate begins to decrease, the curve rises slowly until full saturation is reached.

In the interpretation of a sequence of runs, it should be considered that the absorption run of a fresh solution builds up its own Tris·bicarbonate buffer and its curve pattern will be somewhat different from those that follow a regeneration step.

Figure 8 shows the performance of 1 liter of 0.04 M Tris solution containing 0.01% enzyme in five consecutive runs over 96 hrs. Figure 9 represents the performance of 1 liter 0.08 M Tris solution containing 0.01% enzyme over a period of 33 hr. The feed gas rate in both cases was 500 ml./min., with 1% CO$_2$ by volume. The absorption temperature was kept at 20°C., and the solutions were regenerated in vacuum at 30°C. Both solutions absorbed more

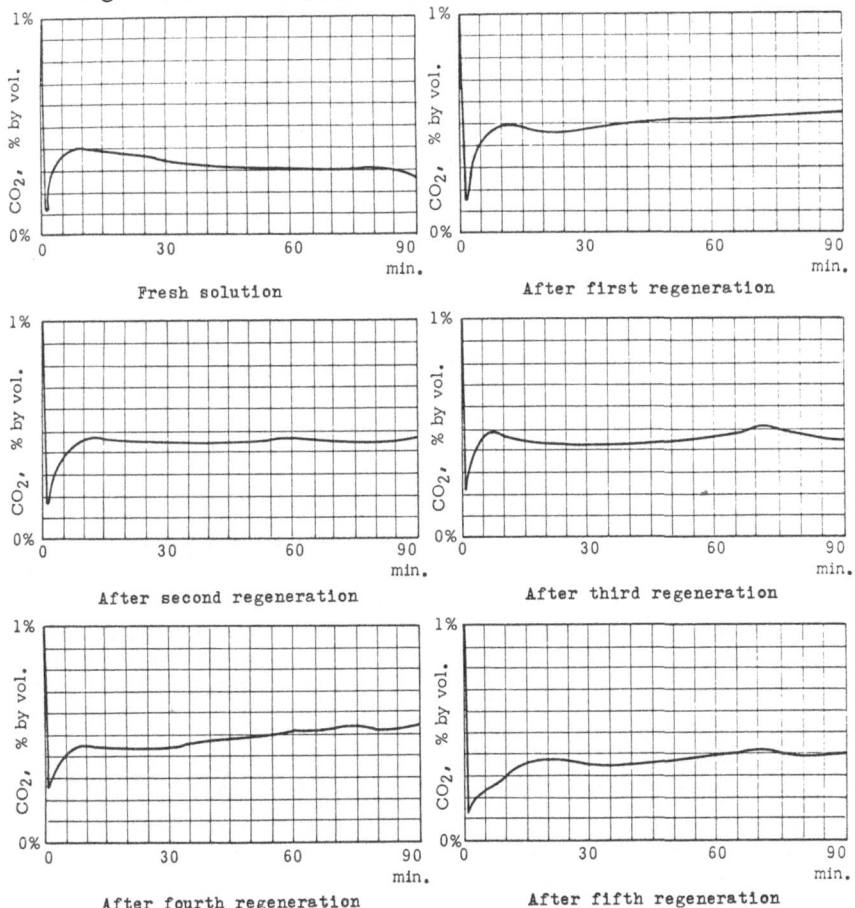

Figure 9. CO$_2$ absorption in cycling operation. Absorber: 0.01% CA in 0.08 M Tris, 1.000 ml.; data are recorded in same sequence as in Figure 8.

than 0.5% CO_2 from the feed gas. The steady performance after repeated regeneration shows no signs of enzyme inactivation.

Table 1 presents data obtained by evaluating the absorption curves of Figure 8. Table 2 presents similar data based on runs represented in Figure 9.

The purpose of these experiments was to prove the feasibility of a cycling operation and no precautions were taken to increase

TABLE 1

CO_2 ABSORPTION DATA IN CYCLING OPERATION[a]

Description	CO_2 entering absorber (mmoles)	CO_2 absorbed (mmoles)	CO_2 absorbed (mmoles/min.)	CO_2 absorbed (%)
Fresh solution	18.0	12.0	0.134	66.8
After:				
First regeneration	18.0	9.0	0.100	50.1
Second regeneration	17.98	9.8	0.109	54.6
Third regeneration	17.76	9.6	0.107	54.2
Fourth regeneration	17.74	9.2	0.102	51.9
Fifth regeneration	17.91	11.4	0.126	63.5

[a] Absorber: 1,000 ml. 0.01% CA in 0.04 M Tris; feed gas: 1% CO_2 in air; feed gas rate: 500 cc./min.; temperature: 20° C.; absorption time: 90 min.

TABLE 2

CO_2 ABSORPTION DATA IN CYCLING OPERATION[a]

Description	CO_2 entering absorber (mmoles)	CO_2 absorbed (mmoles)	CO_2 absorbed (mmoles/min.)	CO_2 absorbed (%)
Fresh solution	17.81	13.3	0.148	74.8
After:				
First regeneration	17.82	11.6	0.129	65.1
Second regeneration	17.74	12.0	0.134	67.9
Third regeneration	17.55	11.4	0.127	65.1
Fourth regeneration	17.47	11.4	0.127	65.3
Fifth regeneration	17.62	11.4	0.127	64.7

[a] Absorber: 1,000 ml. 0.01% CA in 0.08 M Tris; feed gas: 1% CO_2 in air; feed gas rate: 500 cc./min.; temperature: 20°C.; absorption time: 90 min.

the absorber efficiency by improving the rate of interphase transfer of CO$_2$ by better gas distribution or by packing the absorber.

In order to test the capacity of the system by changing "inner" variables, such as enzyme and Tris concentrations, the feed gas rate was increased, and first the enzyme concentration and then the enzyme and buffer concentrations were raised. Increasing the enzyme concentration did not improve the absorber performance; the combined effect of higher buffer and enzyme concentrations resulted in an excellent absorption pattern, as illustrated in Figure 10.

The desorption pattern was improved by increased enzyme

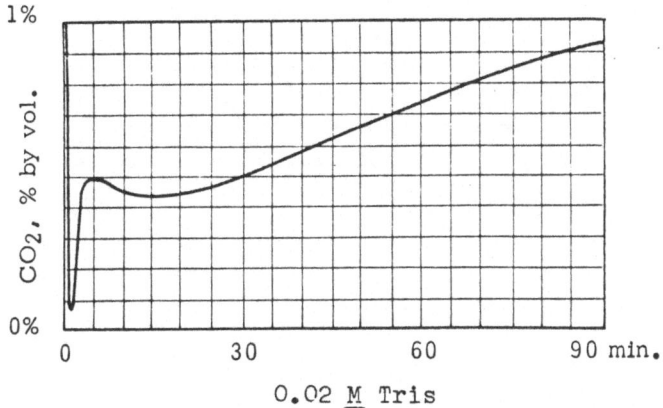

0.02 M Tris

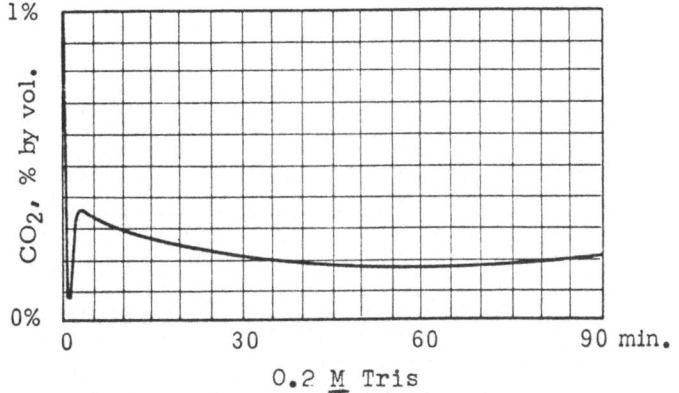

0.2 M Tris

Figure 10. Effect of buffer concentration on CO$_2$ absorption at high feed gas rate. Absorber: 0.02% CA in 0.02 and 0.2 M Tris, ml.; feed gas: 1% CO$_2$ in air; feed rate: 1,000 cc./min.; temperature: 20°c. Recorded percent CO$_2$ in effluent.

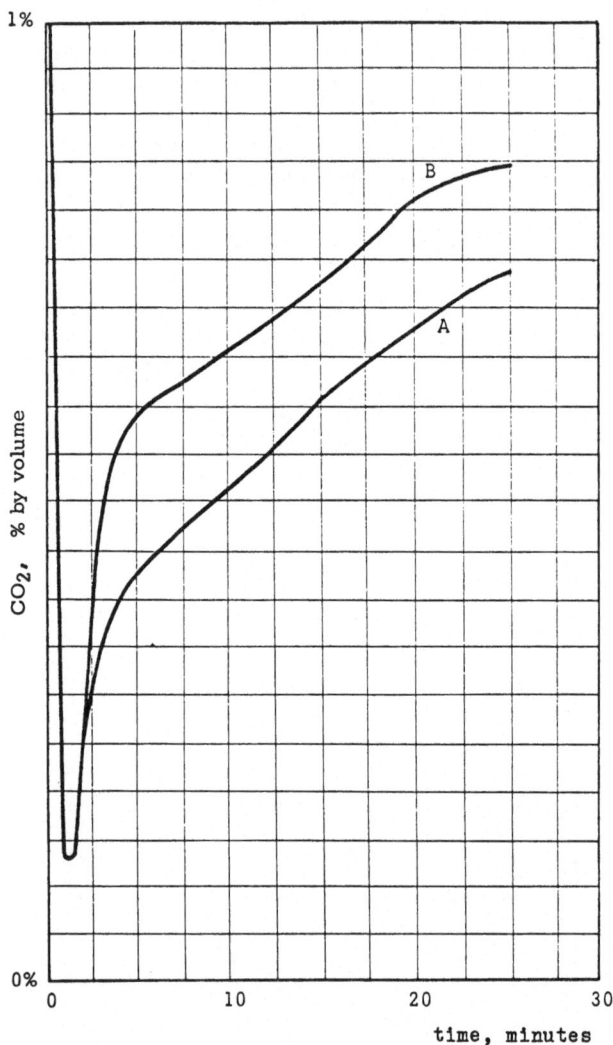

Figure 11.   Effect of carbonic anhydrase concentration on the regeneration efficiency. Absorber: 0.02%
CA in 0.02 M Tris, 1000 ml.; feed gas: 1% CO_2 in air;
feed rate: 1,000 cc./min.; temperature 20°C. Recorded:
percent CO_2 in effluent; A, regenerated with 0.01% CA;
B, regenerated with 0.02% CA.

concentration. Figure 11 shows the results of such an experiment. Two 500-ml. dilute absorbent solutions, each containing 100 mg. CA, were quickly saturated by the feed gas at the rate of 1,000 ml./min. One solution was supplied with an additional 100 mg. enzyme before regeneration. Regeneration time for each solution lasted 10 min. Then the second solution was supplied with 100 mg. more enzyme, so that after regeneration the solutions each contained a total of 200 mg. CA. Thus, at this point, the solutions differed only in the previous history of regeneration. When the solutions were absorbing under identical conditions, the solution that was regenerated in the presence of 200 mg. enzyme showed improved absorption characteristics. This procedure was chosen because of the short time needed to perform the experiment. In a usual long run the absorption pattern may change, due to other factors than enzyme concentration, during regeneration.

LABORATORY PERFORMANCE AND SYSTEM REQUIREMENTS

The performance of this laboratory absorber had to be evaluated on the basis of the reprocessing requirements of closed spacecraft atmospheres.* The CO_2 production, water vapor production, and the oxygen consumption is determined by the metabolic rate of the crew. It is assumed that on the average the metabolic rate of CO_2 is 1,020 gm./man-day and the metabolic oxygen consumption is 907 gm./man-day. The water vapor evolved from respiration and perspiration is approximately 1,000 gm./man-day. Besides processing these major atmospheric constituents, trace contaminant removal is mandatory.

An ideal integrated atmospheric control system, operating in a closed loop, would show the following general arrangement: air from the cabin is circulated through the trace contaminant removal subsystem. The air is then processed in the humidity control subsystem. Part of this dehumidified air is channelled to the CO_2 management subsystem; the remaining air flows directly to the

---

*It is difficult to define an optimum cabin atmosphere for long-duration missions, but it is the consensus that a shirt-sleeve atmosphere is desirable for the crew. Therefore, the cabin temperature should be maintained at 21°C. Cabin pressures of 510 to 520 mm. are presently considered optimum, with oxygen concentration varying from 30% to 45% by volumn. The use of a diluent gas, preferably nitrogen, is almost mandatory because pure oxygen atmospheres constitute a fire hazard. The CO_2 content should not exceed 1% by volume at 760 mm., or 1.5% by volume at 520 mm.

cabin. The CO_2 is removed in the CO_2 management subsystem and then directed to the oxygen-recovering subsystem for processing. The recovered oxygen is then rechannelled to the cabin. The system must also feature a gas storage for balancing and emergency purposes.

The steady-state CO_2 balance is determined by the rate of metabolic CO_2 production of the crew, the flow rate of the cabin atmosphere through the absorber system, and the efficiency of the absorber.

The rate of metabolic CO_2 production is expressed in terms of the respiratory quotient, R.Q:

$$\text{R.Q.} = \frac{\text{moles } CO_2 \text{ produced/day}}{\text{moles } O_2 \text{ consumed/day}} = \frac{\text{vol. } CO_2 \text{ produced/day}}{\text{vol. } O_2 \text{ consumed/day}}$$

If N is the number of crew members and m_{O_2} is the mass flow of O_2 per day (i.e., of O_2 consumed), the number of moles of CO_2 produced by the whole crew during 24 hr. is calculated as $(\text{R.Q.}) (m_{O_2}/32) (N)$. This amount of CO_2 is produced at an approximately constant rate and is being contuously added to the reprocessed atmosphere from which the CO_2 has been totally or partially removed. Since the reprocessed air reenters the cabin at a constant rate, a steady-state level of CO_2 is maintained. This condition is expressed as:

$$(\text{R.Q.}) (m_{O_2}/32)(N) = x^*_{CO_2} \eta_{CO_2} F_r$$

where η_{CO_2} is the efficiency of the reprocessing system, x^* the mole fraction of CO_2 in the cabin atmosphere, steady-state value, and F_r the mole rate of flow of cabin atmosphere through the reprocessing system per day.

Since R.Q., N, η_{CO_2} and m_{O_2} are assumed to have fixed values, and $x^*_{CO_2}$ is also fixed by physiological requirements, the main variable is the flow rate of the cabin atmosphere through the system. (R.Q. and m_{O_2} show physiological variations and η variations of performance, all of which are considered here negligible in first approximation.)

For a one-man crew under regular physiological conditions, R.Q. is 0.82 and m_{O_2} is 907 gm./day. If the steady-state value of CO_2 is to be maintained at a level of approximately 1% by volume ($x^*_{CO_2} = 0.01$) and the efficiency of the reprocessing system is assumed to be 1.0, the mole rate of flow of the cabin atmosphere is calculated as:

$$F_r = \frac{(\text{R.Q.})(m_{O_2})(N)}{(32)(\eta_{CO_2})(x_{CO_2})} = \frac{(0.82)(910)(1)}{(32)(1)(0.01)} = 2331.9 \qquad \text{gm.-moles/day}$$

The average production rate of CO_2 is 1020 gm./man-day, or 0.71 gm./min. or 350 ml./(min-man), at room conditions. Thus the reprocessing system for a one-man crew should absorb CO_2 at this rate. Assuming 0.66 efficiency, the flow rate should be kept at about 50 to 55 st. liter/min. to maintain a steady-state 1% CO_2 concentration by volume ($x^*_{CO_2}$ =0.01).

If the cabin pressure is reduced to 520 mm. Hg, the same amount of CO_2 in the cabin atmosphere requires $x^*_{CO_2}$ =0.015, and the volume flow rate of the cabin atmosphere through the absorber becomes, at η_{CO_2} = 0.66, F_γ vol. = 33.3 liters/min. at 0°C. and 520 mm. Hg.

For room temperature (21°C.) these values have to be corrected. It should be considered, however, that at reduced pressures the respiratory quotient is raised somewhat and the preceding values are somewhat altered.

The laboratory system absorbed CO_2 from an air stream containing 1% CO_2 by volume while passing through the absorber at a rate of 500 ml./min. in intermittent operation. The efficiency of this system could be easily maintained at the level of 66% absorption or above. The air passing through the absorber was maintained at 1 atm. pressure or slightly higher.

Based on the assumptions made earlier, this laboratory system should be scaled up in a ratio of 1:100 to meet the requirement posed by a one-man crew. In case of reduced cabin atmospheres, the recycling rate should be accomplished with a minimum increase in volume and weight penalty imposed by the system. On the basis of the laboratory experiments, it can be safely assumed that this is possible by increasing the buffer and enzyme concentrations simultaneously and by increasing the over-all efficiency of the system in a continuous operation.

Continuous operation

This was attempted on a laboratory scale. Figure 12 is a schematic representation of the apparatus used in these experiments. An absorption column and a desorption column were connected in a closed loop and the liquid, containing buffer and enzyme, was circulated through this system. In the absorption column the liquid came in countercurrent contact with the feed gas; in the desorption column, the liquid came in countercurrent contact with the purge gas. The effluent gas from the absorption column was passed through a gas analyzer and the residual CO_2 was monitored continuously. At present, no precautions have been taken to remove and collect the CO_2 from the purge gas. The liquid was circulated by a peristaltic pump at variable speeds. The feed gas

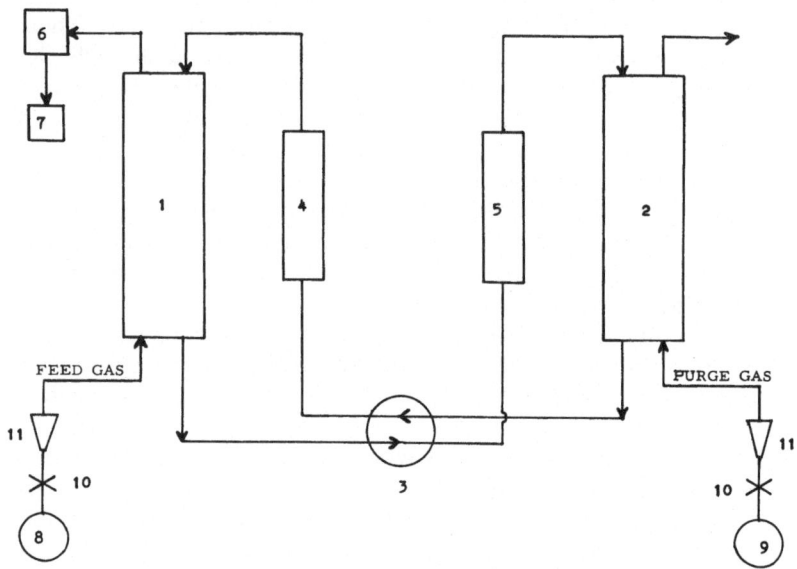

Figure 12.   Schematic of continuous CO_2 absorption apparatus.
(1) absorption column; (2) desorption column; (3) peristaltic pump;
(4) cooler; (5) heater; (6) gas analyzer; (7) recorder; (8) feed gas
cylinder; (9) purge gas cylinder; (10) valve; (11) flow meter.

and the purge gas were dispensed from compressed gas cylinders
and the flow rates were controlled by metering valves and rota-
meters.

The feed gas was a certified mixture of 1% CO_2 in balance air
from Matheson; in these experiments, nitrogen was used as purge
gas. The absorbent solutions were prepared with the same chemicals and the same variety of enzyme as described in the previous
experiments.

In order to maintain a temperature gradient between the absorber and the desorber, the solution was preheated to 30 to 35°C.
by a heating tape before entering the desorption column and it
was cooled to 20°C. before entering the absorber by passing the
liquid through a condenser.

The length of each column was 122 cm. with an I.D. of 5 cm.
The absorption column was packed with glass beads of 5 mm.
diameter, the desorption tower with glass beads of 12.7 mm. diameter. The height of packing in both columns was 84 cm. The
packings were supported by perforated plates above the fritted
glass gas distributors to allow for flooding area. The liquid was
distributed at the top of the column through a shower-type liquid
distributor.

This laboratory system was tested in 14-hr. continuous operation. During this experiment, 1,500 ml. absorbent liquid was circulated through the system at a rate of approximately 500 ml./min. The absorbent solution contained 0.01% CA in 0.5 M Tris. The feed gas, 1% CO$_2$ in air, was passed through the absorber at a rate of 1,500 cc./min. The purge gas rate was varied between 1,000 and 2,000 cc./min. The absorption temperature was kept at 20°C.; the desorption temperature at 30 C. With this arrangement 70% to 80% of the CO$_2$ was absorbed at a steady rate, and changes in the purge gas flow rate had very little effect on the efficiency of the absorption.

In similar but shorter runs, there were indications that the system can operate efficiently at a much narrower temperature gradient at a higher liquid circulation rate. When both columns were kept at 20 C., 70% to 75% absorption was achieved by increasing the liquid circulation rate to 1,000 ml./min.

The removal of CO$_2$ by using a purge gas is attractive provided the CO$_2$ concentration in the purge gas can be increased according to the need. (One of the fundamental problems of CO$_2$ management is to concentrate the CO$_2$ for use in the oxygen regeneration processes. Present regeneration methods require high concentrations.) Schematically, in a continuous operation, coupling an absorber with a desorber, the CO$_2$ is transferred from one gas phase (the feed gas) to another gas phase (the purge gas) through a liquid transducer. The driving force of this transfer between the phases is the partial pressure gradient across the system. Since in these experiments the purge gas was emptied in the atmosphere, the gradient between the feed gas concentration and the atmospheric CO$_2$ concentration was approximately constant during the whole operation. This effect was enhanced by the fact that the pressure drop in the absorber was larger due to the type of packing. The fact that this absorption and desorption is not a mere physical process, but is accompanied by a catalytic reaction, makes the theoretical prediction of rates more complicated (Norman, 1958). It requires the knowledge of the mechanism of the reaction, the rate of the reaction, the diffusivities of the reaction products and equilibrium constants, since the reaction is reversible. In view of the extremely rapid catalytic hydration and dehydration of CO$_2$, it is tempting to assume that the rate of absorption and desorption is controlled entirely by diffusion in the gas phase; however, the transfer of the product from the absorption site to the desorption site must be controlled by the circulation rate of the liquid in addition to the diffusion of the product in the liquid phase.

By changing the ratio of the feed gas rate to the purge gas rate, a more favorable CO$_2$ removal is accomplished for further processing. In the laboratory operation it was shown that a ratio of 4:1 or 5:1 of the two gas flow rates easily allows for a 70% to 75% CO_2 removal from the feed gas. Presently a systematic study is under way in this laboratory to investigate the nature of this enzyme-catalyzed continuous process, and to furnish the data necessary to construct a scaled-up model for testing under simulated conditions. The problem of controlling the liquid-gas interface under zero g is not considered here. This critical problem is related to such questions as liquid propellant handling and may be solved effectively by the surface tension properties of the liquid through proper container geometry (Petrash and Otto, 1964).

STUDY OF A MULTIENZYME SYSTEM FOR CO_2 MANAGEMENT

The use of a multienzyme system was occasioned by M. Utter's recent discovery of pyruvate carboxylase (pyruvate: CO_2 ligase, EC 6.4.1.1), which catalyzes the formation of oxaloacetate from pyruvate by carboxylation (Utter and Keech, 1963; Keech and Utter, 1964).

$$\text{Pyruvate} + CO_2 + \text{ATP} \rightleftharpoons \text{oxaloacetate} + \text{ADP} + \text{Pi}$$

The oxaloacetate formation is essential for the maintenance of the citric acid cycle since decarboxylation of oxaloacetate to pyruvate may occur spontaneously and the decrease of oxaloacetate concentration would decrease the rate at which the citric acid cycle would normally operate. Among the several mechanisms from which oxaloacetate may be regenerated, CO_2 fixation by pyruvate, the process catalyzed by Utter's enzyme, seems to be the most efficient. Essentially, the process represents a shortcut, bypassing the formation of phosphoenolpyruvate as an intermediate between pyruvate and oxaloacetate (Utter *et al.*, 1964). This is suggested mainly by the fact that in the case of the newly discovered enzyme the thermodynamic equilibrium strongly favors the oxaloacetate formation, and the low value of the Michaelis constant seems to suggest that CO_2 fixation may occur at a much lower partial pressure than in other cases. The reaction has an absolute requirement for acetyl CoA which points to the possibility of a positive feedback mechanism. (An eventual increase of the acetyl CoA level increases the demand for oxaloacetate, the formation of which in turn goes on at a higher rate by the enhanced activity of pyruvate carboxylase through acetyl CoA itself.) The enzyme also contains biotin as its prosthetic group which results in a ternary complex mechanism, a characteristic of CO_2 ligases of subgroup 6.4. The energy requirement of the reaction is supplied by breaking the high energy phosphate bond of ATP.

The CO_2 fixation by pyruvate can be reversed by reduction of oxaloacetate to malate with malate dehydrogenase (L-malate:NADP oxidoreductase, EC 1.1.1.37), and subsequent oxidative decarboxylation of malate by "malic" enzyme*[L-malate:NADP oxidoreductase (decarboxylating, EC 1.1.1.40) (Straub, 1942; Graves *et al.*, 1956; Ochoa et al., 1948).

$$CO_2 + pyruvate \rightleftharpoons oxaloacetate$$
$$\searrow \quad malate \quad \nearrow$$

Pyruvate carboxylase was prepared according to Utter's procedure in partially purified form from lyophilized chicken liver mitochondria with a specific activity of 20 enzyme units/mg. protein. Malic enzyme was contained in this partially purified preparation, as tested by its NADP specific activity. Malate dehydrogenase was commercially available.

The CO_2 exchange by this multienzyme system was tested in a Summerson manometer. (Essentially this system is organized by specificity. The soluble enzymes are mixed in a solution with complete mutual accessibility; see Dixon and Webb, 1964.) The enzymes and their coenzymes were suspended in Tris solution. (A 100-ml. solution was made up of 1.2114 gm. Tris, 0.1111 gm. sodium pyruvate, 0.1558 gm. ATP, 0.0049 gm. MgCl₂· 6 H O., 0.0100 gm. acetyl CoA, 0.1680 gm. NaHCO₃, 0.0180 gm. reduced NAD, and 0.79 ml. malate dehydrogenase solution.) The system was filled with a standardized gas mixture of 1% CO_2 by volume in air at atmospheric pressure. The absorption was initiated by adding the pyruvate carboxylase to the reaction mixture from the side-arm of the manometric flask, and the absorption was monitored by manometric measurements and by observing the change in absorbance at 340 mμ. The desorption was effected by addition of NADP. In the initial cycle 86% of the CO_2 was recovered, the difference being accounted for by the formation of Tris · bicarbonate buffer.

These preliminary experiments suggest that such a multienzyme system can be used for reversible CO_2 exchange between a gas and liquid phase. It should be noted, however, that the experiments were carried out on a semimicro scale under static

---

*"Malic" enzyme is an older trivial name for the enzyme in question. Presently it is not recommended since it covered a variety of enzymatic activities. The accepted trivial name is malate dehydrogenase (decarboxylating) (NADP).

conditions, and that no attempt was made to improve the enzyme stability or to rephosphorylate ATP. The significance of these primary investigations lies in the prospect that the NAD/reduced-NAD or NADP/reduced-NADP system is probably accessible to electrochemical control such that the ATP requirement can be kept at a minimum and the substance carried as a high energy "fuel" with minimum weight penalty. Although pyruvate carboxylase is quite unstable under preparational conditions and also appears to be cold labile, it can be stabilized by storing in sucrose solution. Probably another more efficient method of stabilization would be to insolubilize the enzyme according to the method of E. Katchalski (1960; 1961). By this method enzymes could be converted into water-insoluble products with retention of specific activity. Preparation of columns with enzymatic activity could be attempted and the insolubilized enzyme or a multilayered enzyme system could be used for a column-type operation similar to a natural environment in which enzymes are tied to the cristae of mitochondria. (Further development along these lines may open a new route of biochemical engineering, with the eventual result of complete atmospheric control. This would be an organization of enzymatic activities by structural factors, not merely by chemical specificity.)

SUMMARY

The full development of a solid-state multienzyme system in column-type operation will require the coordinated effort of many laboratories and scientists of various fields; thus, this problem is beyond the potential of our laboratory. On the other hand, the second line of our investigations on the feasibility of enzymatically controlled CO_2 management systems proved to be much more accessible. The results obtained by using carbonic anhydrase for CO_2 management are briefly summarized in the following:

It was feasible to use carbonic anhydrase (carbonate hydro-lyase, EC 4.2.1.1) for reversible CO_2 absorption at high feed gas rates, and at low partial pressures. The absorbing liquid contained at least 0.005% to 0.02% enzyme in dilute Tris solutions.

Carbon dioxide absorption and desorption by this system was alternated in cycles over periods of 7 days. Each cycle consisted of a 90-min. absorption and a 60-min. desorption phase. The activity of the enzyme was not changed during these cycling operations. The breakthrough time of these intermittent operations was extended by increasing Tris concentration.

The effectiveness of the cycling operation can be increased by shortening the desorption phase. This can be achieved by higher enzyme concentrations.

Increased enzyme and buffer concentrations enhance the efficiency of the absorbing phase. In intermittent operation, 0.66 to 0.70 efficiencies are easily maintained.

The cycling operation was replaced by a continuous operation in which one leg of the loop absorbed CO_2 from the air-feed and the other leg desorbed CO_2 into a purge gas-feed. The continuous operation required less volume of absorbing liquid to achieve the same efficiency as did the intermittent operation. The efficiency of the absorption could be maintained at 0.75 to 0.80. The activity of the enzyme was not imparied during the continuous operation.

Acknowledgments

This project was sponsored by the Research and Technology Division, Wright-Patterson Air Force Base, Ohio, under Project No. 6146, Task No. 614611. The monitor of the program was Dr. J. P. Allen of AF Flight Dynamics Laboratory, who suggested the biochemical approach for CO_2 management in space capsule atmospheres.

We are grateful to Dr. D. Burk of National Institutes of Health, Dr. S. Ochoa of New York University, Dr. M. F. Utter of Western Reserve University, and Mr. Louis Kovach, Director of Research, Barnebey-Cheney Co., for their friendly interest and most valuable suggestions. We acknowledge the participation of James Gray, Daniel W. Handel, James Schrode, and Hartmuth Schroeder, students of Youngstown University.

REFERENCES

Burk, D. (1961). Ann. N.Y. Acad. Sci. 92: 372.

Datta, P. K., and T. H. Shepard (1959). Arch. Biochem. Biophys. 79: 136.

Dixon, M., and E. C. Webb (1964). Enzymes, 2nd ed., p. 538 Academic Press, New York.

Enzyme nomenclature: recommendations (1964) of the International Union of Biochemistry (1965). (Elsevier, Amsterdam, London, New York.

Gomori, G. (1955). In Methods of enzymology (S. P. Colowick and N. O. Kaplan, eds.), Vol. 1, p. 138. Academic Press New York.

Graves, J. L., B. Vennesland, M. F. Utter, and R. J. Pennington (1956). J. Biol. Chem., 223: 551.

Katchalski, E. (1960). Nature, 188: 856.

———, (1961) J. Biol. Chem., 236: 1720.

Keech, D. B., and M. F. Utter (1964). J. Biol. Chem., 238: 2609.

Keilin, D., and T. Mann (1940). Biochem. J., 34: 1163.

Lindskog, S., and P. O. Nyman (1964). Biochim. Biophys. Acta, 85: 462.

Meldrum, N. U., and F. J. W. Roughton (1933). J. Physiol. 80: 113.

Nahas, G. G., ed (1961). Ann. N.Y. Acad. Sci., 92: 333.

Norman, W. S. (1958). In Chemical engineering practice (T. Davies, ed.) Vol. 6, p. 67. Academic Press, New York.

Nyman, P. O., and S. Lindskog (1964). Biochim. Biophys. Acta, 85: 141.

Ochoa, S., A. H. Mehler, and A. Kornberg (1948). J. Biol. Chem. 174: 979.

Petrash, D. A., and E. W. Otto (1964). Astronaut. Aeronaut., Vol. 2, p. 56.

Report of the Commission on Enzymes of the International Union of Biochemistry (1961). Pergamon Press, Oxford.

Roughton, F. J. W. (1961). Biochem. J., 43: 550.

———, and J. Booth (1946). Biochem. J., 40: 319.

———, and A. M. Clark (1951). In The enzymes, 1st ed. (J. B. Sumner and K. Myrbäck, eds.), Academic Press, New York. Vol. I, Pt. 2, 1250.

Tashian, R. E., D. P. Douglas, and YaShion L. Yu (1964). Biochem. Biophys. Res. Communs., 14: 256.

Utter, M. F., and D. B. Keech (1963). J. Biol. Chem., 238: 2603.

———, D. B. Keech, and M. C. Scrutton (1964). In Advances in enzyme regulation (George Weber, ed.), Vol. 2, p. 49. Macmillan, New York.

11.

THERE'S NO PLACE LIKE SPOME

ISAAC ASIMOV

Department of Biochemistry,

Boston University School of Medicine, Boston

A FREE EXERCISE IN SPECULATION

Let me begin by coining an uneuphonious word—spome—and de-fining it.

A spome is any system, substantially closed with respect to matter, that is capable of supporting human life for an indefinitely long period of time.

The Earth is a spome and, at present, is the only spome known to exist. Its qualifications for spomehood are obvious. It has sup-ported human life for well over a million years, if we count the hominids generally, and will continue to do so for the forseeable future, barring the effects of man's own willful folly.

Furthermore, it is substantially closed with respect to matter. The matter that is added in the form of meteoroid infall or lost in the form of atmospheric leakage is not significant. It does not af-fect Earth's spomic characteristics, nor is it likely to in the for-seeable future.

But a spome cannot be closed with respect to energy.

Life is a process whereby relatively unorganized components of the environment are made more organized. That means that life involves a continuing decrease of entropy and can exist only at the expense of a continuing, and even greater, increase of en-tropy in the environment generally.

If the Earth were closed with respect to energy, mankind, and life generally, would see to it that in a relatively short time, enough oxygen and organic matter would be degraded to carbon dioxide and other wastes to render the Earth uninhabitable.

The energy of the Sun makes all the difference. It enters the Earth-system, keeps the atmosphere stirred up and the oceans liquid; it makes the rain fall; most important, solar energy is utilized by green plants to reconvert carbon dioxide and water into organic substances and free oxygen.

The entropy of the environment, pushed upward by the

activities of life, is pushed downward again by the energy of the Sun. An equilibrium has been maintained for some billions of years at the expense of the vastly increasing entropy of the Sun.

(Beyond the Sun we need not go. For all we know, there are processes that reverse the entropy increase of the stars and keep the universe in stable equilibrium forever, as some astronomers believe, but that need not concern us. The Sun will endure, substantially in its present form, for some ten billion years and that on the human scale, is, an indefinitely long period of time. Earth may therefore be regarded as a spome.)

If the Earth were the only spome that could exist, the subject of spomology would be trivial. It would be comprehended by such sciences as geography and geology. But it may be that the Earth is merely the only spome that exists so far, and that many others can exist in conception or potentiality. In that case, the subject increases in interest.

It is possible, of course—indeed, it is certain—that elsewhere among the stars (but *not* in our own solar system) there may be other spomes. There may be planets sufficiently like the Earth in general characteristics, with a sun sufficiently like our own Sun, to serve as habitable planets. This possibility has been analyzed most completely by Stephen H. Dole of Rand Corporation in his book "Habitable Planets for Man" (Blaisdell, 1964). Without repeating his analysis, I will state his conclusion that in our own single galaxy, the Milky Way galaxy, there are perhaps 640,000,000 habitable planets very much like Earth.

In other words, 640,000,000 potential spomes.

And yet all 640,000,000 lumped together do not in themselves suffice to make spomology a truly interesting study, for they are all merely so many Earths. From the broad standpoint of the spomologist, if you see one earthlike planet, you have seen them all. Since we have all indeed seen one earthlike planet, our own, we have seen them all and can forget about them.

What we want, if we are to make spomology interesting, are spomes that are drastically different from the Earth. And if we make the subject interesting, we may find—who knows—that it is valuable as well.

Suppose we ask ourselves what makes Earth a spome and Jupiter or Mercury non-spomes? If we want to express the difference most succinctly, it is a matter of mass. Jupiter is too massive, Mercury is insufficiently massive. The difference in mass involves, one way or another, almost every quality that goes to make or not make a spome.

If a planet is insufficiently massive it cannot hold either an atmosphere or an ocean of a volatile liquid. If it is too massive, it will hold hydrogen and helium and produce a poisonous atmosphere and, at best, an ammoniated ocean. In neither case can it be a spome.

If it is very massive, that is probably because it is far distant from its primary and can accumulate matter with little competition from the greater body, and at a temperature low enough to make the dancing molecules of hydrogen (the major component of matter) sufficiently sluggish to be captured. Under such conditions, the planet is too cold to be a spome.

If it is insufficiently massive, it is because it is too close to the primary, so that accumulating matter is lost to the greater body and many of the more common elements are, at that distance from the primary, too nimble and elusive to be captured. Alternately, the body is forming too close to a large planet which competes successfully for matter, so that the body itself is a satellite rather than a planet. In the former case, the body is too hot to be a spome, in the latter too cold.

There are exceptions to these rules, of course; known exceptions within our solar system. Our Moon seems too large for its place in the system, whereas Pluto seems too small. This departure from regularity leads to theories that the Moon is a captured planet and Pluto an escaped satellite.

On the other hand, assuming a sun of the proper type, it is quite reasonable to hope that there is a good chance that a planet of the proper mass would be bound to form at the right distance from that sun and with the proper chemical composition to lead to spomehood.

We might even say that the search for a spome is the search for a body of appropriate mass.

But all this is in the course of nature. It works if we are looking for "natural spomes," for spomes ready-made. Let us now add the factor of human intelligence. Only God may make a tree, according to Joyce Kilmer, but perhaps spomes can be made by fools like us. (No, I didn't invent the word in order to be able to make that statement.)

The problem is: Can we make an "artificial spome"? Can we take a body of drastically wrong mass and make a spome of it? In one direction, let's not even try. Bodies too massive to be spomes are quite rare (there are only five in the solar system, counting the Sun itself). They are, in addition, too dangerous to play with, thanks to their strong gravitational fields and their inevitably enormous atmospheres.

On the other hand, bodies insufficiently massive for spomehood are common. There are thousands of them in the solar system. In fact, the closest body to us, the Moon, is an example of the class.

The problem boils down to the conversion of small bodies into spomes, and the specific version of the problem is, inevitably: Can we make the Moon into a spome?

The Moon is certainly not a spome now. Thanks to its low mass, it has neither an atmosphere nor free water. But let us consider essentials and not accidentals: An atmosphere can be kept from diffusing out into space by the force of a sufficiently strong gravitational field, but, on a smaller scale, it can be kept from doing so by physical barriers as well.

In other words, we can distinguish two general varieties of spomes: external and internal. An external spome is one with an atmosphere and ocean held to the outer surface of the body by a gravitational field, so that men can live on that outer surface. An internal spome is one with air and water held within an airtight cavity and with men living on the inner surface. Inevitably, natural spomes are external ones, while artificial spomes must be internal.

Suppose, then, we hollow out a cavity under the Moon's surface and supply it with air, water, and the other necessities of life. We might have to begin with capital from the Earth, but it is possible that eventually water could be baked out of silicate hydrates in the body of the Moon. From such water, oxygen could be formed.

Given a sufficient supply of energy, and a mass of variegated chemical composition such as the Moon (or even a much smaller body), the basic chemical requirements can be met on the spot.

Energy is the key, and we are used to thinking of the Sun as the energy source. In nature, the only source of energy in quantities large enough to support a natural spome does, in fact, happen to be a star like our Sun, but a star—any star—is an incredibly wasteful source. Hardly any of its radiation is stopped by a planet, and only a small fraction of that which is stopped is used. A much smaller source, used with much greater efficiency, will serve the purpose.

A roaring wood fire, whose energy production is a completely contemptible fraction of that of the Sun, will warm us in winter at a time when all the Sun is insufficient. On the scale sufficient for internal spomehood, however, we will need something considerably better; and that something better is in sight.

On the large scale of spomehood, only hydrogen fusion can be relied on as an energy source through an indefinite future. It is large-scale hydrogen fusion that powers the Sun, and it may be small-scale hydrogen fusion that will power the Earth some day.

I foresee, then—although not in the immediate future—the possibility of the Moon being honeycombed immediately below its surface by a growing system of caverns, supplied with all basic materials from the Moon itself, with all its energy requirements supplied by fusion power plants, seeded with plant and animal life (and, inevitably, with microscopic life as well), and inhabited by men, women, and children; families who may know no other life, and want none.

The advantages are obvious. The Moon will have a controlled environment designed specifically for man; man will have what he wants and needs (in many vital respects) and not merely what he can get. What's more, it will have the advantage of a fresh start. As the United States managed to prosper and flourish partly because it was freed of many of the choking traditions of Europe's bitter past, so the Moon, it may be hoped, will be freed of the incubus of Earth's past mistakes.

Some disadvantages are also obvious. However confidently we rely on scientific and technological advance, it seems certain that we can never do anything to alter the Moon's gravity. The inhabitants of the Moon will be at a gravity one-sixth that of the Earth.

Undoubtedly, they can get used to it, and people born on the Moon, knowing no other, will consider such a gravitational force natural. Will men suffer as a result, however, particularly in the transitional period when they may be shuttling between the Earth and the Moon? Will muscles weakened and bones softened under the influence of lower gravity be able to withstand a return to Earth?

The problem might not arise in fullest intensity. Men on the Moon could keep in condition with exercise or in centrifuges. Perhaps only a few specialists would need to condition themselves for possible trips to Earth, whereas the general population of the Moon would find it no hardship to remain away from the Earth.

Another disadvantage is that an internal spome is liable to accidental catastrophes of a sort to which external spomes are immune. An atmosphere and ocean held to the surface by gravity are absolutely secure. Barring catastrophe on an astronomic scale, nothing can alter the gravitational force and nothing can cause the atmosphere and ocean of an external spome to be lost.

On an internal spome, on the other hand, a cavern punctured by a large meteorite, or broken by a landslide, loses its air at once and its water more slowly. Nevertheless it is to be expected that men will be ingenious enough to minimize the chances of such catastrophes. Furthermore, the cavern of an internal spome will undoubtedly be compartmentalized so that a local catastrophe can be confined to its immediate neighborhood.

Nor is catastrophe a bar to spomehood. There are catastrophes on Earth, too. We suffer periodically from the effects of hurricanes, blizzards, tornadoes, floods, and drought, to none of which the Moon would be subject. A patriotic Moonman might well argue that it was the Earth rather than the Moon that fell short of ideal spomehood through catastrophe.

But what about the psychological difficulties? Can men really learn to live for extended periods in what is essentially, after all, a cavern? Can he bear to be born and to die there? The answer in my opinion, is the heartiest possible affirmative. If the cavern is large and comfortable, why not?

It is a mistake to underestimate the flexibility of mankind. Man has already demonstrated abilities to make enormous adjustments. A city such as New York represents, in a way, almost as artificial a spome, one almost as divorced from man's original environment, as the Moon would be. Yet man has made the transition from hut to skyscraper over an insignificant period of time. Indeed, a peasant immigrant can adjust adequately to New York in his own lifetime.

Why should we imagine a Moonman would be horrified at being "cooped up"? I think it would be much more likely that he would think with horror of a world like the Earth, where men had to cling precariously to an outer surface, exposed to the vagaries of an unpredictable and changeable climate. A Moonman might no more want to live on Earth than a New Yorker would want to live on a farm.

Of course, in thinking of an internal spome, we must fight our prejudices. It is easy to fall into the trap of thinking, vaguely, that an external spome is "natural" and an internal spome "artificial," and that what is natural is good and what is artificial is bad.

The argument might even be advanced that a "true" spome can only be one in which life could develop spontaneously out of nonliving matter, as it did on Earth. A world that had to be engineered and seeded by a species that already had two billion years of evolution behind it might seem no true spome at all, but one that was only able to imitate spomehood through an initially parasitic dependence on a true spome.

But if that argument is advanced, where does *Homo sapiens* stand? Life did not develop on dry land. The only portion of the Earth that is a "natural" spome, in the sense that life arose there spontaneously from simple chemicals, is the ocean. It was only little by little that certain types of living things emerged onto the dry land, a habitat as hostile to the creatures of the sea then as the Moon seems to us now.

Some fishy philosopher, if we can imagine one, might well have shaken his head at the foolish creatures who chose to emerge on land. It would seem a bad exchange to move from the equable environment of the ocean to the violent extremes of the open air;

from a plenitude of water to the perennial threat of desiccation; from a gravitation-free three-dimensional world to a gravity-ridden two-dimensional world.

Nor are these dangers unrealistic ones, or these disadvantages of the dry land imaginary. Life first invaded the land some 425,000,000 years ago, yet even today, the ocean remains much richer in life than dry land, area for area. Land animals had to evolve for millions of years before they could develop limbs strong enough to lift them clear of the ground and make both size and rapid movement simultaneously possible. It was some two hundred million years before creatures evolved with internal thermostats and external insulation so that the equable temperature of the ocean might be imperfectly restored. Man himself rose to his hind feet a million and a half years ago and still pays his respects to gravity with flat feet, slipped disks, sinus trouble, potbellies, and numerous other ailments. And to this day he must live in dread of falling, a dread we are usually unaware of only because we are so accustomed to it.

No, no, if we are going to sneer at the Moon as an unnatural habitat, we must sneer with precisely equal intensity at the continents of the Earth. We live on a portion of the Earth aritificially seeded from the truly spomic portion; and despite everything, land life remains less rich and, in some respects and by some criteria, less confortable and less successful than ocean life.

Yet need we be sorry that our ancestors emerged from sea to land? With all land's dangers and discomforts, it opened the way to advances not possible in the sea. In hindsight, we can see that the ocean was a dead end, whereas land offered a new and brighter horizon.

Nor are we being parochial when we argue in this way. Air is far less viscous than water. In water, a creature must either travel slowly or it must be streamlined. The most highly developed sea creatures, the squids, sharks, and fish, are highly streamlined. The land creatures that return to the sea are streamlined in proportion to the extent to which they have returned, if you think of the otter, penguin, seal, sea cow, and finally, the whale.

A streamlined body implies short, stubby appendages, if any, with an exception for the squid's highly specialized tentacles. In low viscosity air, on the other hand, it is possible to be fast-moving and irregularly shaped at the same time, so that land animals can have elaborate appendages. It is to this that man owes his priceless hands.

Consider how, were the porpoise indeed as intelligent as man, the lack of hands would hamper the exhibiting of that intelligence!

If we ever learn to communicate with porpoises we may find ourselves with fluked philosophers on our hands; introverts who can
think but not do.

Then, too, one can deal with fire only in air and never in water.
Only a land creature, therefore, could conceivably develop the technology that begins with the discovery of fire. It is certainly possible
to argue that man's advancing technology is not an unalloyed good,
but I doubt that even the most inveterate yearner after the good old
days before the building of Blake's "dark, satanic mills" could
possibly wish to retreat to the days before the discovery of fire.

To use a chemical analogy, the passage from sea to land involved a "phase change" in the progress of life; one that most or
even all of us cannot help but consider desirable.

Is it possible, then, that the passage from an external "natural"
spome, to an internal "artificial" spome might likewise involve a
desirable phase change? I hate to undertake the role of prophet
here; foresight in such matters is as difficult as hindsight is easy.
Nevertheless, I will try.

It seems to me, for instance, that however difficult the initial
passage from an external spome to an internal one, the end would
be a partial cancellation of the difficulties introduced by the previous
great life-adventure. In an internal spome, man would return to
the equable environment and lower gravity of the sea, without aban-
doning the low-viscosity environment of the air. An internal spome
would have, after a fashion, the best of both land and sea and the
worst of neither. Surely something great may come of that.

If we begin with an internal spome on the Moon, victory and
success there can only inspire attempts at expansion, at forming
spomes out of other medium-sized bodies such as Mars and the
larger satellites of Jupiter. In particular, though, there may be a
movement to internal spomes on smaller and smaller bodies—the
asteroids that exist by the thousand in the space between Mars and
Jupiter.

Why the asteroids?

Well, consider the matter of efficiency. With the best will in
the world, and with all the technological advances likely in the foreseeable future, it would seem that mankind could not burrow very
deeply into the skin of the Earth, or into the skin of even a smaller
body such as Mars or the Moon. We may sink narrow bores to the
mantle in time to come, but if we are thinking of interal spomes, of
large, comfortable, and well-appointed caverns, the outer couple
of miles is the most that we may consider. (Earth's internal heat,
perhaps that of Mars and the Moon too, would make deeper caverns
uncomfortable anyway.)

This means that virtually all the volume of a planet is unused and serves the men of the spome only by supplying them with the source of a gravitational field.

The asteroids, however, can be spomified completely. They can be riddled and honeycombed. They have no internal heat for discomfort and no significant gravity to make more difficult the shifting of mass. Nor need the caverns by buttressed more than minimally to counter possible collapse. If we except the very largest, *all* of an asteroid can be used. (A nickel-iron asteroid might be difficult to work with, and its composition might not be suitable as a source of raw material for anything except the ferrous metals but, judging by the ratio of iron meteorites to stony ones, we can hope that less than 10% of the asteroids will be metallic, the rest stony.

Nor need an asteroid be considered too small to make an ample spome. Some years ago, I wrote a story about such an asteroidal spome, in which an Earthman visiting the asteroid expressed surprise that the inhabitants had room to grow tobacco. His guide to the asteroid replied:

We are not a small world, Dr. Lamorak; you judge us by two-dimensional standards. The surface area of Elsevere [the asteroid] is only 3/4 that of the State of New York, but that's irrelevant. Remember, we can occupy, if we wish, the entire interior of Elsevere. A sphere of 50 miles radius has a volume of well over half a million cubic miles. If all of Elsevere were occupied by levels 50 feet apart, the total surface area within the planetoid would be 56,000,000 square miles and that is equal to the total land area of Earth. And none of these square miles, doctor, would be unproductive.

In the story I deliberately dismissed one serious problem that would inevitably arise on an asteroidal spome in order that I might concentrate on the sociological point I was trying to make. I avoided any consideration of the fact that the gravitational field on an asteroid is negligible by supplying my storybook spome with artificial gravity.

In real life, as opposed to science fiction, an artificial gravity field cannot be set up merely with a wave of the typewriter. One conceivable possibility would be to set the asteroidal spome into rapid rotation. The centrifugal effect would be analogous to a gravitational field directed outward, with some important side effects. If you visualized such a spome made up of spherical levels nested like the layers of an onion, the intensity of the centrifugal effect would depend on the distance from the center. It would be highest in the outermost level and would decrease steadily as one went "upward," becoming zero at the center. In addition, Coriolis forces would be many times more noticeable on a small sphere like and asteroid than on a large one like the Earth. The smaller the spome, the greater the angular velocity required for a given maximum centrifugal effect, and the more pronounced the variations in the effect and in the obtrusiveness of the Coriolis forces.

It seems to me that spinning the spome would not be worth the energy expended and the problems produced. Why not, instead, accept null gravity as a condition of life? Life has, in the past, switched from the essential null gravity of the oceans to the gravity slavery of the land and survived. Why not the switch back?

To be sure, the switch from null-g to g was made over eons of time, and the bodies of the creatures making the switch had to undergo elaborate and glacially slow changes through the force of natural selection. Mankind obviously lacks the time to proceed in this fashion.

But it is not in space science and engineering alone that mankind is experiencing great advances in technology. Biology is undergoing its own revolutionary breakthroughs. It is reasonable to hope that by the time man reaches the point where he can reach the asteroids with a supply of energy sufficient to set up a spome, he will also have learned enough about genetics to engage in meaningful tissue-engineering. Why may we not suppose that the changes necessary to fit a human body for null gravity can be guided by intelligence rather than left to the colossal blindness of a nature that knows only random change?

A null-gravity body may well be designed differently from our own, but not necessarily radically so. Bones and muscles may be smaller and legs shorter, but I would guess that this would not go to extremes. To whatever extent weight may disappear, the body will still have to handle inertial mass, which would be the same on an asteroid as on the Earth.

A null-gravity body would, it seems to me, become utterly graceful in its maneuverings, gaining some of the three-dimensional skills of the fish and birds. We will have a human species capable of flight without having to sacrifice the infinitely useful hand for the sake of a wing.

Land animals might require similar adaptations but, except perhaps for pets, dwellers on the asteroidal spomes could do without them. Plants could be grown at null-g without very much trouble. Fish could still be cultivated. Algae culture and the chemical industry might combine to produce food items with the taste and texture of meat if that were desired.

To be sure, a null-g man can never come to Earth, or even visit a world as small as the Moon, but that should be no more a hardship to him than the fact that we can no longer breathe under water is to us (unless we are drowning.)

If we concentrate on this state of affairs, it would seem that there would be two species of man, g and null-g. We are g, of course, as would be the colonists on such large spomes as Mars, the Moon, the large satellites of Jupiter, and so on. The inhabitants of the asteroidal spomes would be null-g.

Might it not be that the passage from g to null-g is the new phase change? That the future will belong to the null-g? That we g's of earth will now reach a dead end, while the null-g's of the asteroids will find a new and glorious horizon opening up for them? They may advance, leaving the discarded things of Earth behind them while we, no more able to follow them than a fish could us, remain as oblivious as fish to their greater glories.

Consider—first, the null-g species may well outnumber us as time goes on. Honeycombed asteroids may support a larger population, all taken together, than the mere outer skin of the large spomes inhabited by g's. The fact that null-g might be smaller in body (though not in brain) would serve to make their possible numbers still greater.

Second, the nature of the null-g environment will make it certain that they will far outstrip us in variability and versatility. The g people will exist as one large glob (Earth's population) with small offshoots on Mars, the Moon, and elsewhere, but the null-g will be divided among a thousand or more equal worlds.

The situation will resemble that which once contrasted the Roman civilization with the Greek. The Romans wrought tremendous feats in law and government, in architecture and engineering, in military offense and defense. There was, however, something large, heavy, and inflexible about Roman civilization; it was Rome, wherever it was.

The Greeks, on the other hand, reaching far lesser material heights, had a life and verve in their culture that attracts us even today, across a time lapse of 2,500 years. No other culture ever had the sparkle of that of the Greeks, and part of the reason was that there was no Greece, only a thousand Greek city-states, each with its own government, its own customs, its own form of living, loving, worshiping, and dying. As we look back on the days of Greece, the brilliance of Athens tends to drown out the rest, but each town had something of its own to contribute. The endless variety that resulted gave Greece a glory that nothing before or afterward has been able to match; certainly not our own civilization of humanity-en-masse.

The null-g's may be the Greeks all over again. A thousand worlds, all with a common history and background, and each with its own way of developing and expressing that history and background. The richness of life represented by all the different null-g worlds may far surpass what is developed, by that time, on an Earth rendered smaller and more uniform than ever by technological advance.

A third difference, and the really crucial one, in my opinion, can best be explained if I now turn to the subject of this symposium, space cabins.

In the light of what I have already said, you can see that a space cabin is not exactly a spome, for a spome must be capable of supporting human life indefinitely. It is rather a "spomoid," something that is capable of serving a spome-like function temporarily. It is the object of this discussion to describe advances made (or about to be made) that will serve to extend the life of a spomoid or make it more efficient; in short, to make the spomoid more nearly a spome.

Spomoids have performed notably well on a number of occasions already. Each astronaut has inhabited a spomoid for a period of from fifteen minutes in a suborbital flight to fourteen days in the case of the recent flight of Gemini 7.

It is the obvious intention of the human race to explore the solar system by means of spomoids even before any extraterrestrial spomes are established; and, in fact, even if it turns out that the establishment of extraterrestrial spomes are unfeasible.

We hope to have an astronaut reach the Moon and return; he won't have to stay there long. Even if we establish an astronomic observatory on the Moon, we do not necessarily need to make a spome of the world for the purpose. The observatory can be largely automated, and human attendants can arrive for short tours of duty.

But while the Moon is still in the pre-spome stage, there is nothing to stop astronauts from setting out on voyages to Mars, Ceres, Ganymede. Optimists among us might well suppose that even though no extraterrestrial spome were set up anywhere within the solar system, it would still be possible to explore the system with more and more elaborate spomoids. By stages, we might even reach Pluto.

But there we would have to come to a halt. Beyond Pluto lie the stars, and the distances there involved are so enormous that the techniques that will have sufficed for the solar system will be completely useless to meet the new situation.

To reach even the nearer stars will involve one of three alternatives:

(1) Straightforward flight from here to the nearer stars and back, the time required being anywhere from a generation to a century or more.

(2) Flight at velocities near that of light, thus introducing a time dilatation effect so that the duration of the flight will seem to the astronaut to be no more than a few months or years. In that case, however, on returning to Earth, he will find that the time lapse on Earth has been anywhere from a generation to a century or more.

(3) Flight with astronauts frozen into suspended animation, the effect being the same as in case 2.

None of these alternatives is pleasing. The astronaut will either have to expose himself to the perils and uncertainties of freezing over long periods of time, or be willing to expend the energies required to reach extremely high velocities. It may well prove that freezing for decades is unfeasible and that the energy demands for time dilatation are prohibitive. If alternative 1 is chosen as the simplest, the astronaut must not only spend most or all his life on the starship; he may also have to be prepared to bring up children and grandchildren who will in turn have to take over the starship and spend their lives on it.

As for those who wait on Earth, there are no alternatives. A starship leaving for a neighboring star may not get back for a hundred years. The original astronauts may shorten the time for themselves by time dilatation, or by freezing, and return scarcely aged, but that does not affect the observers at home. The starship will still not have returned for a century and no one in the crowd that waves good-bye will be in the crowd that waves hello.

Under the circumstances, stellar exploration would never be a popular exercise for anyone, either on the ship or at home. A few expeditions may set off as tours de force, but Earthmen, unable to follow them, unable to see the results in their own lifetime, will lose interest.

But let's consider under what conditions such voyages might become popular.

The longer the exploring trip within the solar system, the more elaborate the spomoid (or space cabin, remember) will have to be. By the time the outermost planets are reached, space voyages will have become years in length and a spomoid capable of supporting a crew for years will, of necessity, have a recycling mechanism that would require little further sophistication to serve a crew indefinitely.

The trend in space exploration, then, will be from the spomoid to the spome and, certainly, where stellar exploration is concerned, nothing less than an elaborate spome will be required.

Not only is a starship a spome, but it is an internal spome; and an internal spome of an extreme type. In assembling a crew for a starship, we are asking Earthmen and women to make the transfer from an external spome to an extremely internal one and we may be asking too much.

To be sure, I have been talking about the establishment of spomes all through this article—*but by stages*! The change from the external spome of the Earth to an internal spome on the Moon is, in many ways, a mild one. There is still the chance of communication with Earth, the sight of the Earth in the sky, even if only on a television set within the cavern, and, finally, the possibility of returning to Earth some day.

It is then the men of the Moon, accustomed to an internal spome, who will go on to spomify Mars and Ganymede. And it will be the far distant colonists, further divorced from the Earth by the mere fact that it is not forever hanging in the sky like a large balloon, who will make the further step to the asteroids and the g-species.

Little by little the inhabitants of spomes would get over any longing for blue skies, open air, the stretch of ocean, the intricate world of mountains, rivers, and animals.

But even a colonist from the Moon or Mars would not feel at home on a starship, which would be null-g, unless it were rapidly rotated—with all the problems that would introduce.

No, the proper crew for a starship would be null-g people, and there would be no need to recruit them, for an asteroidal spome would be a starship in itself. Working upward from a primitive space cabin and downward from the Earth, we meet in the middle at the equation: asteroidal spome = starship.

Now, under such conditions, a voyage to the stars could be made without hardships whatever. If an asteroid were fitted with rocket motors and made to veer out of its course and away from the Sun (the escape velocity from the Sun is considerably less in the asteroid belt than it is in Earth's vicinity) what would it matter to the null-g inhabitants of the asteroid?

They had always been in a null-g internal spome, and they would still be in a null-g internal spome. They wouldn't be leaving home; they would be taking home with them. What matter how long the trip to a star? How many generations live and die? There would be no change in their way of life.

To be sure, they would leave the Sun, but what of that? A dweller of the asteroids would not depend on the Sun for anything. Properly space-suited, he might emerge from the asteroid and observe the Sun as a tiny, glowing marble in the sky, but nothing more. He may miss that sight and idealize "the Sun of home" but such idealizations will evoke nothing more than a nostalgic thought, like the modern city dweller's occasional sigh for the "old home town."

The starship turning out of its orbit might simply be taking the third and final step in the weaning of life. Once species of life were weaned from the ocean. With the establishment of extraterrestrial spomes, life will be weaned from the Earth. With the starships, it will have been weaned from the solar system.

But why should the asteroids bother to become starships? What do they gain? A number of things:

First, the satisfaction of curiosity—the basic, itching desire to know. Why not see what the universe looks like? What's out there anyway?

Second, the desire for freedom—why circle the Sun uselessly forever, when you can take your place as an independent portion of the universe, bound to no star?

Third, the usefulness of knowledge—since a trip of this sort is bound to add to the information possessed and this new information will surely add to the security and comfort of the spome.

Nor need such a journey be dull and uneventful. True, it may take hundreds or even thousands of years to reach a star, and generations may live without seeing one at close quarters, but does this mean there is nothing at all to see?

I can't really guess what phenomena would await the ship and what beauties of nature they will find to admire. One thing seems certain, however; the Universe must be better populated than would appear.

We see the stars because they advertise themselves so brilliantly, but small stars are far more numerous than large ones, and dim stars far more numerous than bright ones. Surely bodies that are so small and dim that they can't be seen, except at close quarters indeed, are the most numerous of all.

Perhaps no generation will pass without some dark world coming into view; some material body the starship may pause to investigate. If the body is large, the starship couldn't land, but it could still fly by, take up a temporary orbit, observe, and nose it out. If the body were small enough to have a negligible gravity, it could be mined and made to serve as a source of minerals to replace the small inevitable losses suffered by any spome, however efficient the cycling.

When the neighborhood of a star was reached, with its lighted planets, observations might be particularly intense and particularly interesting. The system may contain external spomes: earthlike planets bearing life—even, perhaps, intelligent life.

What a rare phenomenon that would be in terms of human lifetimes! How fortunate the generation granted such a sight: Silently, they would observe, watch, and eventually, pass on as the unbearably attractive lure of open space beckoned—and back on the inhabited planet, creatures might talk excitedly of flying saucers.

The neighborhood of a star might offer a chance for refueling, too. I can conceive that the deuterium supplies needed for the fusion reactors might be picked up in the space the ship passes through, but such deuterium is spread out incredibly thinly. It would be more concentrated within a stellar system. The neighborhood of a star might then be not only a means of seeing a rare sight, but also a chance to stock up on deuterium—enough to last another million years or so.

If an asteroidal belt were encountered about some star, a landfall might, in a sense, be made. The starship could take up some appropriate orbit. Other asteroids could then be made into spomes. The colony would divide and new ones would be set up. Eventually one or more of them—or all of them—would set off as starships themselves. Perhaps an old, old starship, worn past the worthwhileness of repair, can be abandoned on such occasions—undoubtedly with much more trauma than ever the Sun and Earth were abandoned.

Little by little as the years passed and lengthened into centuries, millennia, and hundreds of millennia, the starships would swarm in greater and greater numbers, with all the Universe their home.

And every once in a while, perhaps, two spomes would meet by arrangement.

That, I imagine, would involve a ritual of incomparable importance. There would be no flash-by with a hail and farewell. The spomes, having contacted each other in a deliberate search over vast distances, would be brought to a stand relative to each other and preparations would be made for a long stay.

Each would have compiled its own records, which it could make available to the other. There would be descriptions by each of sectors of space never visited by the other. New theories and novel interpretations of old ones would be expounded. Literature and works of art could be exchanged; differences in custom explained.

Most of all there would be the opportunity for a cross-flow of genes. An exchange of population (either temporary or permanent) might be an inevitable accomplishment of any such meeting.

And yet it may happen that such cross-flows will become impossible in an increasing number of cases. Long isolation may allow the development of varieties that may no longer be interfertile. The meetings of spomes will have to endure long enough, certainly, for a check on whether the two populations are compatible. If not, intellectual cross-fertilization will have been carried on, at any rate.

Eventually, perhaps, space will carry a load of innumerable varieties of null-g intelligences, all alike in that space is their home (and, indeed, "space-home" is what the shortened "spome" stands for); in that they are intelligent; and in that they are descended from the inhabitants of some planet that may no longer exist in their memory even as a component of legend, and from which the initial load of humanity may long since have vanished.

It may even be that *Homo sapiens* may not be the only species to make the transition to a starship culture. Perhaps there is a crucial point, reached by every intelligence, from which two roads branch off, one leading to the true conquest of space and the other to a slow withering on the planetary vine.

Out there, perhaps, are many creatures waiting for man to join them. And when we do, we may find ourselves united with them not in terms of material body resemblances, but in the life we lead and in the intellect we cultivate.

Is this, then, the phase change that space exploration makes possible? Or am I only stumbling in a vain attempt to see the unseeable? Perhaps the essential point of the phase change is as far beyond my grasp as the smell of a rose is beyond the grasp of a fish or a Beethoven symphony beyond the grasp of a chimpanzee.

—But I tried.

267